WEIRDNESS!

WEIRDNESS!

WHAT FAKE SCIENCE AND THE PARANORMAL
TELL US ABOUT THE NATURE OF SCIENCE

Taner Edis

PITCHSTONE PUBLISHING
DURHAM, NORTH CAROLINA

Pitchstone Publishing
Durham, North Carolina
www.pitchstonebooks.com

10 9 8 7 6 5 4 3 2 1

Library of Congress Cataloging-in-Publication Data

Names: Edis, Taner, 1967- author.
Title: Weirdness! : what fake science and the paranormal tell us about the
    nature of science / Taner Edis.
Description: Durham, North Carolina : Pitchstone Publishing, [2021] |
    Includes bibliographical references and index. | Summary: "Explains how
    science works and why science is essential by exploring many of the
    beliefs embraced by large sections of the public that are rejected by
    the scientific mainstream"— Provided by publisher.
Identifiers: LCCN 2021028120 (print) | LCCN 2021028121 (ebook) | ISBN
    9781634312110 (paperback) | ISBN 9781634312127 (ebook)
Subjects: LCSH: Pseudoscience. | Science—Popular works.
Classification: LCC Q172.5.P77 E35 2021 (print) | LCC Q172.5.P77 (ebook)
    | DDC 500—dc23
LC record available at https://lccn.loc.gov/2021028120
LC ebook record available at https://lccn.loc.gov/2021028121

CONTENTS

PHYSICS AND WEIRDNESS

Unlike those of my colleagues, the sign outside my office door does not tell visitors that I am part of the physics department. Instead, under my name, I have "Physics and Weirdness."

Those who step into my office might not notice anything unusual at first: a mess that resolves into a desk, a computer, papers, books, and a whiteboard with equations and drawings. But if they look closer at my bookshelves, they will find that my books on physics are outnumbered by those about aliens, psychic powers, creationism, and alternative medicine. I have more books that promote distortions of quantum mechanics than books that help me teach quantum mechanics—more that explore beliefs about a hollow earth and UFOs than texts on proper cosmology.

I also have a lot of philosophy of science and skeptical criticisms of fake science mixed in, to lend the bookshelves with weirdness an air of respectability. I often find myself explaining that studying weird and disreputable claims is an excellent way to approach important questions about the nature of science. If I spend half a day reading websites promoting conspiracy theories, well, I have an excuse: I am doing research about the erosion of trust in scientific institutions. While I started out as a normal physicist, nowadays I work on the history and philosophy of science, with an emphasis on weirdness. I am fortunate enough to teach at a university where I can get away with all this. I am doubly fortunate that I regularly teach a course called "Weird Science," where students debate weird claims and find out that there's a lot more to science than absorbing the content in the textbooks.

I should add that I find weirdness fascinating. Claims that challenge established science, particularly if they come from outside mainstream institutions, grab my attention. I will even warn colleagues that if I start to get very interested in what they do, it may be because I think it's weird. If a paranormal or fake scientific claim also has a sizeable popular following, that's even better. Almost always, I think, the weird claims are implausible: blatantly false, sometimes crazy, at best resident in a kind of gray zone where I can't yet label them completely wrong but they still seem very fishy. But what we get wrong about how the world works is, I think, as interesting as what we get right.

So that's what this book is about. I want to explore what the disreputable fringes of science might tell us about real science. I will bring up denial of evolution or climate change, opposition to vaccines, paranormal and supernatural claims, Bigfoot, and conspiracy theories about aliens or secret societies. But I won't linger too long on the details of why each form of weirdness fails to describe our world correctly. What I am most interested in is what the weirdness means for science. I spend much of my time helping students to learn particular aspects of physics. But science is not just a body of knowledge; it is also a process of learning, an institution entangled in politics, and a significant cultural presence. Yes, if a creationist thinks that evolution is a myth, she makes a mistake. But examining that mistake can tell us a lot about more scientific ways of understanding our world. If an anti-vaccine parent spreads conspiracy theories about the medical establishment on social media, he can undermine public health as well as propagate falsehoods. But such examples only highlight questions about when we should trust scientific institutions. After all, we typically don't have full expertise about a matter on hand, and we have to start by asking whom to trust. When it comes to vaccines, for example, my background in physics is not much of an advantage—I also have to figure out how much to trust health-related institutions.

In the following pages, I will tell some stories—I have some interesting ones, after many years of teaching Weird Science and poking around the fringes of science. I will throw in a few weird parables. I think weirdness is fun; I hope that some of my fascination with weird claims will come through. I will bring up what I think are the most interesting questions, and when I feel confident, propose some answers.

The Plan

I will start with the way that scientists and skeptics often compose checklists detailing the features of proper scientific method, such as falsifiability. Skeptics then argue that weird claims do not have all these required properties. This is a philosophy of science that is outdated by decades. I will argue that instead, we should learn from science studies and start to focus on how institutions can promote or fail to promote learning about the world.

Paranormal and fake scientific claims often violate the laws of nature. Understanding what such a violation is, and why it offers strong evidence against a claim, requires looking at modern physics and how physical laws are connected to symmetries. Some weird claims purport to reveal facts beyond physics, perhaps beyond any science. I will discuss quantum physics and the scope of combinations of rules and randomness, showing how implausible such claims are.

Supernatural beliefs within religious traditions and claims of paranormal ways of knowing can strongly challenge scientists and skeptics, since debates don't just turn on evidence but introduce legitimate concerns about trust and community. Evolution is a common flashpoint. I examine the popular models of the relationship between science and religion, including harmony, conflict, or separate spheres. None of these do full justice to the complex questions raised by supernatural convictions.

What we learn from skepticism about paranormal claims can also help us criticize and improve more established forms of knowledge. The study of weirdness illuminates questions about metaphysics, mathematics, applied science, social science, and the humanities. Old-fashioned ideas about the unity of science that specially favor natural science do not work, but I argue that our various knowledge-producing disciplines can still productively support and offer criticism for one another.

Skeptics and scientists typically account for weird beliefs through a combination of cognitive biases and reasoning errors. I argue, however, that sometimes even rational belief-forming mechanisms can lead to an affirmation of false beliefs. In a highly nationalist environment, for example, many people may be better off believing some false historical narratives. Rationality is more complicated and more context-dependent than skeptics and scientists often appreciate.

Finally, I observe that today, we confront science denial in a political environment shaped by right-wing populists who distrust expertise, while existential threats such as our climate crisis hang over our heads. This is an occasion to examine the institutions promoted by right-wing populists, the pressures on current scientific institutions, and the association of science with liberal meritocratic politics. I find plenty to criticize even in political stances that are seemingly friendly to science.

The Blind Scientists and the Elephant

There once were six scientists who were blind, and who had never heard of an elephant. One day, one of their students entered their lab and announced that an unusual new creature, an elephant, had been brought to campus. Being curious, the scientists went down to investigate.

The first scientist felt the elephant's leg and said, "Hey, it's like a pillar." She then stood back and observed that "with that thickness, and because muscle strength scales like area—length squared—and mass scales like volume—length cubed—the elephant must be a large and heavy animal. I estimate a mass of five to ten tons and a height around three meters; though I'd need to go back to the office to do a better calculation."

The second scientist got her hands on the elephant's tail. She observed that "the hairs indicate that this must be a mammal. Therefore, if this is a natural beast and not some escaped genetic experiment, I would expect four of these thick legs you just mentioned, and a head on the opposite side of the tail—we should feel around there next."

The third scientist felt the trunk. "Whoa!" she said, "here's something I didn't expect. It seems like it might be a kind of nose extension from the nostrils at the end, but I haven't encountered anything like it before. In fact, let's feel around more carefully to make sure this is an organic part of the body and not a prank our students have played on us."

The fourth scientist had been touching the ear, and she said, "These flaps might serve as a cooling mechanism. That bore Voinovich has some theories about heat exchange mechanisms in large animals. I'd hate to think he was correct, but we should get in touch with him."

The fifth scientist, running her hand over the elephant's belly, called to her first colleague, and pointed out that if a student were to fetch a measuring tape, they could quickly improve on the initial size estimates.

The sixth scientist, who had been handling the tusk, exclaimed, "Here's another surprise! With this hardness and the location on the head, I can only guess that it must be a hugely overgrown tooth. Could it be for digging? Or defense? Sexual display? Hey, can somebody go to the back and tell me if we have a female or a male?"

After some more time investigating, the six blind scientists went back to their lab. After a few days, they put together a short report describing the elephant, and they even got their 3D printer to produce a small model. Giving the model to a student, one of the scientists explained, "Those of you who can perceive electromagnetic waves should know that the orange stripes we painted on it are just a guess. But we have some new equipment coming in that should be able to do scattering off large objects and imaging with electromagnetic waves; if we have access to an elephant then we should be able to make a more confident guess about the color scheme."

Meanwhile, a local mystic grumbled that it was just like scientists to ruin a perfectly serviceable parable.

The elephant just wanted to escape captivity.

1 BEYOND THE EVIDENCE

The Spiritualists Summon Darwin

A few years ago, I had a student in my Weird Science class who came from a Spiritualist family, immersed in the now-fading religious subculture of séances and spirit mediums. She was wonderful to have in the classroom when we were examining claims of psychic powers. But that does not make up a whole semester: we also found our way to discussing creationism and evolution, and my students had to write about what they thought of creationist arguments.

My Spiritualist student turned in a paper describing her childhood experiences with the spirits. Her father was once stationed in Italy, and down in the basement of their building, they saw a vision of a group of Roman soldiers marching by. The floor, however, cut the soldiers off at their knees: they were marching on an old Roman road still buried under the building. And then there was her mother relating a tender memory of watching my student's grandfather holding her—a few months after he had died.

This was certainly more interesting than a common student paper on creationism, but as I read it, I got increasingly concerned that I'd have to give it a failing grade for being completely off-topic. In her last pages, she described a full-blown séance. Someone in her family got curious and tried to settle the creation-evolution dispute once and for all. They would summon the spirit of Charles Darwin, who, with the benefit of being situated on the Other Side, could lay the matter to rest. The spirit of Darwin appeared and gave an answer in line with Spiritualist doctrine: evolution was correct insofar as it described the progression of how our material bodies came to be, but it was incomplete. Biological evolution needed to be complemented by an account of spiritual evolution to give the full story. And with that evidence, my student declared that the creationists were wrong and the evolutionists partially right.

That was the most unconventional defense of evolution I have ever encountered. A few years later, I started telling the story to my Weird Science class. I usually steer the discussion toward two questions.

First, was my Spiritualist student's argument *scientific*? Almost no one who hears the story thinks so. I then ask my students to be more generous in their conception of science. After all, isn't there more than a hint of empirical investigation in the séance the Spiritualists conducted?[1] Sure, Spiritualist beliefs are full of the miraculous and supernatural, and most of us would be inclined to file them under religion rather than science. But my Spiritualist student was not asking her readers to trust in a tradition or to take a leap of faith. The séance was supposed to be a concrete demonstration producing evidence available to everyone, much like what happens in a science lab.

My students may entertain the possibility that a séance might resemble a scientific experiment, but they invariably think it's too much of a stretch. Science involves things like math, fancy equipment, and lots of introverted people wearing eyeglasses. Sitting in a darkened room summoning the spirit of Darwin is completely out of place. It just doesn't fit.

The second question is whether any of this is *true*. Spiritualists are convinced that communication with the dead is real. They think that spirits can give us reliable information about our world, not just the Other Side. Perhaps séances are too different from labwork, and so it may be useful to maintain a distinction between science and talking with spirits.

But whatever the conventions of our language, surely what matters is whether Spiritualists more-or-less have it right. In fact, a main reason we care about whether a practice is scientific is that most of us think that science is a usually reliable process that typically gives us an accurate handle on our world. Once we acknowledge something as science, we also extend our trust. That does not, however, mean that we should only trust science. If séances are not scientific, then maybe Spiritualists need to work a bit harder to convince outsiders, but they still could be correct.

When confronting the question of truth, my students are often torn. On one hand, Spiritualist claims look like religion, and it seems socially abrasive to question religious beliefs. On the other hand, Spiritualism is not mainstream, and my students are predisposed not to accept unfamiliar religious claims. If they're more secular-minded, students may think that a weird practice such as a séance cannot reliably produce truths. If they're religious believers, they may think that mainstream science is not capable of properly investigating spiritual realities, but still be suspicious of occult practices. And there are always a few who evade the question by asserting that a claim can be true for some people and not others.

While my students usually have a skeptical initial reaction to Spiritualism, their skepticism is superficial. Both the question of distinguishing science from rivals and pretenders, and the question of how that is relevant to truth, turn out to be a lot knottier than they first might have thought.

Checklists for Science

One attractive approach to our two questions is to emphasize the process of doing science. Science, we might say, is much more than inert knowledge encased in textbooks. It is a very particular sort of investigation, defined by its method rather than its results. The practice of holding a séance does not comport with the scientific method, and claims like communicating with the spirits of the dead have not been validated through the scientific method.

If we could pin down a scientific method better than the vague secondary-school recipe of inventing a hypothesis, testing it in the lab, refining and rinsing-and-repeating, that would be very useful. A checklist or a set of diagnostic criteria to separate genuine science from pretenders and fakes would have immense practical value. For our health, do we trust mainstream medicine with its invasive procedures and drugs, or alternative practices that seem gentler and more holistic? Do we think that epidemiologists who trace the progress of a pandemic are acting according to the scientific method, or do we suspect that they have been co-opted by governments intent on social control? A good set of diagnostic criteria could help legal systems more easily sort out what was sound scientific expert testimony and what was bunk. An easy-to-use checklist could help us determine what deserves to be part of a common knowledge base, and what are more sectarian articles of faith.

I first came across such diagnostic criteria when I found out about skeptics dedicated to criticism of paranormal and fake scientific claims. Some scientists and philosophers have always worried about the popularity of beliefs such as psychic powers, creationism, and astrology, especially when the weirdness attains media prominence and begins to overshadow mainstream science. In the 1970s, a group of American academics joined forces with a prominent magician or two, who compared what they saw as the dishonest trickery of psychics and Spiritualists with the honest deception of conjuring as entertainment. They formed the Committee for the Scientific Investigation of Claims of the Paranormal (CSICOP, now CSI), and began publishing a magazine, *The Skeptical Inquirer*.[2] I read the back issues in the library, subscribed, and I have been hooked ever since. There is now an international movement of skeptics combining intellectual criticism of weird claims with a consumer-protection orientation toward enterprises such as alternative medicine.

Naturally, a checklist to separate real from fake science appeals to skeptics. Some of the diagnostic criteria advanced in the early days of CSICOP drew from mid-twentieth-century philosophy of science. The virtues of good science included simplicity; following Occam's Razor, it was best to have the simplest accounts making the fewest assumptions. In contrast, paranormal enthusiasts were content to either compile lists of The Unexplained or gratuitously invent otherworldly entities to explain spooky experiences. The best science produced repeatable experiments, while psychical research did not. Skeptics compiled a growing list of items to help diagnose whether an idea was good science or a kind of intellectual pathology.[3]

Sometimes fake science directly challenges mainstream science. Religious conservatives in the United States have long wanted to insert what they call "scientific creationism" into public education. This means periodic lawsuits, where a checklist can come to the rescue. Perhaps science is naturalistic in its method: it is limited to investigating natural causes for natural phenomena. After all, how else can we isolate a system and conduct controlled experiments? Creationists claim that a supernatural agent is responsible for the forms of life; therefore, if science is defined by natural explanations, then creationist ideas are automatically out of bounds and cannot be part of public science education. Such diagnostic criteria have contributed to winning lawsuits against creationists.[4]

The most prominent item in the checklist has been *falsifiability*. Many skeptics and scientists have thought that the possibility of being shown to be false was the distinguishing feature of scientific claims. Scientists need to test their ideas, which means that they take the risk that the results of an experiment will find that their claims are false. In contrast, creation-scientists who assert that the Earth is just thousands of years old, when confronted with the evidence of radiometric dating, make up an excuse. Noah's Flood, for example, is supposed to have been a supernatural intervention that makes all extrapolations from present physics invalid.[5] Being committed to a religious doctrine, creationists cannot risk admitting that their beliefs are false. According to the falsificationism championed by philosopher Karl Popper, science is about making risky conjectures.[6] Indeed, we should not speak of our best theories as being more or less correct, only that they have so far survived a battery of potentially falsifying tests.

When skeptics and scientists need to distinguish real science from fake science, or just when they need to indicate increasing scientific rigor, they often call on falsifiability. As part of my efforts to permanently postpone growing up, I like to keep an eye on new discoveries about dinosaurs. In one book I was recently reading, the paleontologist wanted to communicate how advanced his science had become in the last decades, so much so that we have accurate scientific knowledge of even the colors of some dinosaurs. He started with a summary of falsificationism, citing Popper's work in the 1930s, and went on to point out how with today's improved equipment and field methods, paleontologists can more precisely test ideas rather than rely on informed guesswork.[7]

Criteria such as falsifiability have shaped what has now become a stereotyped rhetoric about how weird claims are excluded by the scientific method. For example, since Bigfoot is supposed to be an unscientific claim, skeptics may be reluctant to say that there is no such animal. Instead, in the pages of a recent *Skeptical Inquirer*, I read that "millions and thousands of years ago, there was an animal that was Bigfoot-like, but there is no evidence that any similar creature roams the earth today. Under the framework of the scientific method, we are able to disprove those explanations not consistent with evidence. As great as this method is, we cannot disprove the existence of something we have no evidence for, in this case, Bigfoot. We can merely say that there is no evidence to support its existence."[8] We might have thought it safe to conclude that Bigfoot was not real, but scientific method dictates that "no evidence" is the best verdict we can have. After all, what is unfalsifiable cannot also be known to be false.

In a world full of weird claims and a social media environment awash with conspiracy theories, it will not do just to ask the public to trust science. Skeptics and scientists have to explain why science is trustworthy. Posting a meme or wearing a t-shirt saying "Science: It Works, Bitches" is not enough—*why* does it work, and what is so special about *science*? We cannot just rely on the track record of our sciences, since that would use our alleged empirical knowledge to validate our empirical knowledge: a circular argument. We need something stronger than just poking at the world, a secure foundation not subject to the contingency of facts about the world. And our answer very often is the scientific method. When applicable—when investigating natural phenomena—our method certifies that we produce reliable knowledge. Scientists can screw up like anyone else, fudging their data or misusing their instruments, so science is fallible. But there is a logic of figuring out how the natural world works, and this logic is expressed by the scientific method. There is a clear difference between science and weirdness, and checklists of diagnostic criteria pick out the distinctive logical features of scientific reasoning.

The Madness in the Method

I am all for keeping creationism out of schools, studying dinosaurs more rigorously, and preventing public discourse from being derailed by conspiracy theories. If anyone advances such causes by ritual invocations of scientific method, I won't complain too much. But if we want to use weirdness to learn more about the nature of science, then our checklists and diagnostics are a bit too crude. Sometimes they are even dead wrong.

Let me emphasize again that the philosophy of science adopted by scientists and skeptics mostly comes from the mid-twentieth century—falsificationism, for example, goes back to the 1930s. This is like making an argument that depends on dinosaurs, but taking dinosaurs to be lumbering lizard-like beasts who dragged their tails on the ground, ignoring our more recent scientific picture where dinosaurs are closer to birds, their surviving descendants. In the past decades, historians and sociologists of science have learned a lot more about how science operates, and philosophers of science have similarly made progress. If there is any short summary of their more up-to-date picture of science, however, it is that science is messy and complicated, and that the prospect for finding special logical features that underwrite the advance of science is not good. There are lots of methods that help us learn about the world, and lots of good advice encapsulated in diagnostic checklists, but no scientific method that can guarantee the validity of knowledge in any domain.

The older philosophies favored by scientists and skeptics present a heroic image of science. Armed with the proper methods, scientists wrest knowledge from nature and underwrite the progress of modern times. In much of the twentieth century, the philosophy of science took such a heroic image for granted. But in the 1960s, anti-establishment attitudes caught up with the philosophy of science. Attempts to figure out a logic of science had already run into technical problems and stagnated. Some successors of Popper adopted approaches close to falsificationism, but ended up portraying science as an arbitrary authority. And some philosophers who tried to read the patterns of the history of science ended up suggesting that the overarching conceptual frameworks of science did not and could not change through rational, evidence-based processes. Certainly not all students of science turned hostile toward notions of scientific progress, but the sexy ideas and the Big Names began to be associated with a more cynical view of mainstream science.[9]

In the following decades, as skeptics and scientists continued to worry about public fascination with UFOs and psychics on one hand, and a resurgent conservative religiosity with its creationism and faith-healing on the other, many academics got caught up in what came to be called the Science Wars. The friends of established science faced off against critics embracing the self-devouring postmodernist doubt then fashionable in the humanities, and fireworks ensued.[10] Meanwhile, most of what went on in science studies was far less exciting, building up a more detailed, nuanced understanding of how various sciences operated. But the Science Wars taught skeptics to interpret dissent from a heroic picture of science as a general hostility toward science, and scientists who had occasion to reflect on the nature of science often found it easiest to ignore what was happening on the other side of campus.

It will help to look at some of the reasons why falsificationism has not been a viable philosophy of science for many decades.[11] First of all, unfalsifiability is not, in fact, a feature of weird claims. A statement such as "the Earth is about six thousand years old" is not structurally different from "the Earth is about four and a half billion years old." Scientists have long had multiple, independent lines of evidence indicating that the time scale for the Earth was much longer than thousands of years; after the advent of modern radiometric dating techniques, age estimates converged on the more precise value that is in today's textbooks. In this context, young-earth creationism is almost certainly false.

Similarly, there is nothing inherently wrong with claiming that a very large ape inhabits the forests of the Pacific Northwest. It is just very implausible, for reasons very similar to why I can be confident that there is no triceratops living in my back yard. Bigfoot is supposed to be a large animal with a breeding population. Therefore we should have found physical evidence for Bigfoot, but we have not. We have plenty of poor quality evidence: stories of sightings, anatomically suspicious footprint casts, a low-resolution movie of what might be someone in a gorilla suit. None of this is a strong signal for Bigfoot that stands out from the background noise we can expect from hoaxes, perceptual errors, and all the mistakes that can happen in a complex world and uncontrolled environment.[12] We all are fallible and no one is entitled to absolute certainty. Nonetheless, nobody needs to apologize for judging that Bigfoot is very likely a fiction. Certainly no one should adopt the jargon of pronouncing "no evidence" because it seems more scientific.

What, then, about the way that defenders of weirdness often move the goalposts? Yes, creationists may respond that Noah's Flood reset all clocks, and Bigfoot enthusiasts may turn Bigfoot into an interdimensional being that does not leave the traces a mere large ape would.[13] But these are observations about the evasive behavior of some creationists and Bigfoot supporters—not features of the claims themselves.[14] For that matter, skeptics can easily criticize the refurbished claims. Noah's Flood should have left all sorts of historical and geological evidence, but it has not. In the context of what we know, it is an absurdity.[15] Piling on even more crazy propositions as qualifiers does not improve the standing of the young-earth claim. Compared to a large ape, a dimension-jumping Bigfoot is an even worse explanation of any concrete evidence, and is as credible as an invisible triceratops in my back yard.

Now, evasive behavior may well be a sign of an intellectual pathology. It can be diagnostic of a community more interested in defending a belief central to its identity than learning about the world. By adjusting peripheral beliefs, the community can protect more important core doctrines. The trouble is, the scientific community similarly adjusts claims in the face of unfavorable experimental results.

Consider the problem of dark matter in cosmology. Astronomers can make good estimates of the mass of luminous matter in what they observe. This is just our ordinary, everyday matter, which interacts through electromagnetic forces and therefore can emit and absorb light. At astronomical distances, however, gravity is the only significant force. And when astronomers measure the speeds of rotation within galaxies and clusters of galaxies, they cannot account for their findings through the gravity of the mass observed.[16]

One possibility is that the measurements are wrong. For example, a recent experiment reported particle speeds faster than light, which would violate our current best understanding of physics. The experimenters suspected a fault, and asked for help to locate it—no easy task, with today's extremely complicated equipment. Fortunately, they succeeded in identifying a subtle problem with a cable.[17] But this option is not open to astronomers. Since there are so many independent observations using techniques validated in many different situations, it is hard to doubt the rotation-speed data without also questioning our very ability to make astronomical observations.

Another possibility is to suppose that our mass estimates are incomplete. This is where dark matter comes in. Since our instruments can't detect any light from it, any extra mass in and around the observed galaxies must consist of particles that only interact with ordinary matter through gravity. If we assume that such dark matter exists, we can account for our data, but only if there is a lot more dark matter than ordinary matter, arranged in large, low-density haloes around galaxies. The problem is, there are few other reasons, and certainly no reason from our earthbound knowledge of physics, to cook up dark matter. Without independent evidence for dark matter, it may look like astrophysicists have invented something new, exotic, and invisible, just to avoid acknowledging failure.

Finally, our theory of gravity might be wrong. Some astrophysicists have tried to explain the data with modified theories of gravity, but their proposed changes also have little independent motivation other than to fit the puzzling data. Moreover, their approach has a hard time accounting for both the galaxy and the galaxy-cluster data, so the dark matter conjecture is by far the leading contender in the astrophysical community. Still, modifying the theory is an option.[18]

In such situations, speaking of falsification is not useful. The theory does not fit the data, but what has been falsified? The experimental procedure? Boundary conditions such as claims about the amount of mass present? The underlying theory? And the next steps in trying to understand the data often look like the scientific community is trying to protect core ideas by just making up new stuff.

Skeptics and scientists are already convinced that there is something wrong with creationism or Bigfoot before they ever adopt the rhetoric of falsifiability to suggest that there is a sharp difference between real science and pretenders. Whatever the difference may be, falsificationism does not pick it out.

Science as Statistics

One of the attractions of falsificationism is its austere logical rectitude. Thou shalt not go beyond the evidence! Popper's key insight was supposed to be that it was impossible to be sure that "all crows are black" was true, since it was not possible to examine all crows. But if we saw only one white crow, we could be certain that "all crows are black" is false. If we have seen 643 black crows, that is the extent of our knowledge. We should be impressed that "all crows are black" has survived 643 potentially falsifying tests, and we are entitled to announce that there is no evidence for nonblack crows, but that is all.[19]

Statistical reasoning, which has now spread to almost all our sciences, depends on a different conception of evidence. Say we have a factory producing plush crow toys. The dyeing process, however, is not perfect: occasionally a nonblack crow gets produced. If we want to know how often errors happen, it won't be feasible to inspect every toy. Instead, we can take a random sample from the production run. If the sample size is large enough, but still much smaller than the total production, almost all such samples will be *representative* of the toy population, giving us an excellent idea of the error rate and the color distribution among the toy crows. We might be wrong—we might, for example, have the extreme bad luck that dyeing errors are due to a software bug that results in a fault in the dyeing mechanism for three minutes every Tuesday at 10:26, and we just happened to take our sample at that time. But such scenarios are very unlikely.

If we adopt logical austerity, our evidence will falsify those claims incompatible with the items tested, but say nothing beyond. But if evidence is representative, it will point beyond what we have actually tested, informing us about a larger population. *The whole point of evidence is to go beyond the evidence.*

I often teach physics to biology majors, and while they're working in the lab, I sometimes flip through their non-physics textbooks. First-year biology texts often start with a description of the scientific method. They suggest that research in biology is an exercise in hypothesis-testing: come up with an idea, beat it into shape to formulate a proper hypothesis attached to an appropriate statistical model, go into the lab and try to get a representative sample to test the hypothesis, and see if the prescribed statistical tests support acceptance or rejection of the hypothesis. Depending on the results, rinse-and-repeat.[20]

Curiously, hypothesis-testing can lead to an unstable marriage between statistical reasoning and the rhetoric of falsifiability, where rejection of a hypothesis takes on the role of falsification. Indeed, in large and influential areas of science, such as biomedical research, the idea of a special scientific method is alive and well. Especially beginning students can easily think of science as a fixed set of rules: we take data, feed it into the machinery, turn the crank, and out pops reliable knowledge about the natural world. If this is so, statistics must be the engine that drives science.

Most statistical tools in science come from classical statistics, a motley collection of techniques used largely because long experience tells us they do a good job. A better candidate for a foundation of scientific method is Bayesian statistical inference, which is not very different in practice, but is conceptually much cleaner.[21] There is a long-running philosophical dispute between partisans of classical and Bayesian statistics, but especially with recent advances in machine learning, the Bayesian approach appears to be gaining popularity. I have run into Bayesian ideas not just in philosophy and natural science, but also in psychology, social science—even history. And I will admit to a soft spot for Bayesian statistics. The first paper I ever published used a Bayesian approach in statistical physics.

Bayesians assign probabilities to theories. They start with an initial probability expressing how strong a theory is, and as evidence comes in, use a simple and elegant mathematical formula to adjust the probability upwards or downwards.

Speaking of theories being more or less probable might work, the way astrophysicists judge that dark matter is a much better prospect than modifying gravity. I am less happy with attaching a definite probability to a theory; for example, declaring that the initial probability of dark matter is 89.36%. This seems like false precision—the sort of thing I bark at my biology students about when they come up with unwarranted precise numbers in the physics lab. Vagueness can be appropriate in science, and definite initial probabilities assume too much.[22]

Still, once it gets going, the Bayesian process of adjusting probabilities is compelling. If the evidence coming in is the sort of thing a theory would predict, the probability of the theory goes up; if the observations are more of a surprise, the probability comes down. Bayesians think that with probability, we have the perfect device to capture the shades of gray between the 100% of certain truth and the 0% of utter falsity.

The initial probabilities, however, remain a problem. For example, when testing a product line of toy crows, we can pick a statistical model with initial probabilities that implements the idea of representative sampling, run our Bayesian machinery, and conclude that, say, the nonblack proportion of crows is likely to be around 2%. But with a different model, we can also reproduce falsificationist austerity. We could say that each distinct configuration of black and nonblack colors in the toy population was a different theory, and that since we start from a condition of ignorance, we will assign each configuration the same initial probability. In that case, each individual toy crow observed from our sample will be evidence that falsifies those configurations that do not include the right color. We will then conclude both that we know nothing about the toys that have not yet been observed, and also that, regardless of how many observations we make, that the likelihood of the next individual crow being black is always 50%.[23]

Such a conclusion is absurd. No statistician proposing to test a factory production run would use such a statistical model. And yet, nothing about that model violates the Bayesian machinery. The reason statisticians do not use crazy models is good judgment, which typically includes a lot of tacit background knowledge about processes like factory production. Any picture of statistics as an apparatus which takes in data and spits out knowledge ignores the work that goes into choosing an appropriate model. This is particularly clear in physics. Sophisticated statistical error analysis is routine in experimental physics, but its success depends on background knowledge about instruments and experience with the kinds of errors that occur in labwork. Giant particle physics experiments could not function without automated statistical analysis to identify signals of novelty that stand out from the background noise. But the statistical process depends critically on the theoretical background provided by our knowledge of fundamental physics.

Statistics is a very useful, broadly applicable tool available to scientists. It is not the only tool, and it is not an overarching method. When scientists try to improve their disciplines, they often demand more statistical rigor. But recurrent methodological disputes and calls for increased statistical vigilance are most characteristic of younger sciences where there may be plenty of data, but also a lack of strong theoretical frameworks that can help structure the data. In a more mature science such as physics, there is less of a need to display methodological credentials. For example, I have not yet come across any physics textbook that describes a scientific method. Indeed, the culture of physics encourages a degree of skepticism about predetermined methods. I like to quote Percy Bridgman, a Nobel laureate in physics, who said that "there is no scientific method as such," merely scientists doing the utmost with their minds, "no holds barred."[24] So in physics, we treat statistical machinery as one tool among many. Many physicists spend careers without using much statistics or doing any hypothesis-testing, particularly if they are theorists or builders of instruments.

None of this means that good statistical practice is optional. In some of our newer sciences, particularly in the biomedical and behavioral disciplines, a lot of published research is of poor quality. Even supposedly established results vanish when other investigators test them, and scientists worry about a crisis of replication. In many fields, the convention for statistical significance is such that about one in twenty reported results would have occurred without any real reason, just due to a statistical fluctuation. On top of that, there are many opportunities for bad practice, such as checking data for multiple possible effects until finding something, which increases the chance of spurious results. It might, as some scientific leaders and journal editors suggest, help to tighten statistical methods, and perhaps even learn from the convention for significance in fundamental physics, where the chance of a spurious effect has to be less than one in about three and a half million.[25]

Tighter statistical methods, however, do not cure all ills. Mucking around with data in the absence of a solid theoretical background will always have limitations. For example, testing the effects of a drug, even with the best statistical practice, cannot substitute for figuring out the extremely complex biochemical interactions in varied environments that underlie the effects of the drug. Sometimes such a theoretical background is just not available. In such cases, statistics can be one of the very few good tools at hand. But it is not the engine of science, or a universal method for learning from data.

Science Without Foundations

In the late 1980s, I started graduate school. What became the internet started taking shape. I subscribed to *The Skeptical Inquirer*. Skirmishes in the Science Wars broke out. And I joined an email list on BITNET called SKEPTIC, intended for discussion and criticism of paranormal beliefs and fake science. When the list needed a new home, I had my university host it and took over managing it.

I also started spending more time with philosophy books than was good for a physics graduate student. Discussions on the SKEPTIC list pushed me toward the philosophy of science. I was never convinced by falsificationism, but otherwise, I was drawn to the quest to establish a distinct logic of science. I was already inclined to think that science gave us the closest approximation to truth available, which had contributed to my shift from engineering to physics as an undergraduate. My experience doing physics in graduate school was only strengthening my convictions. But there was more to it. I grew up in a nonreligious household, at a time when religious politics was rising and secular political options were collapsing onto a cramped individualism. If I looked for something that was universal and positive beyond the narrow pursuit of personal gain, well, it had to be science. I was good at math, bookish, and if I could contribute a small amount to the grand adventure of human progress, well, I'd do it. Science was a quasi-moral cause as much as an excuse to indulge my curiosity.

So my motives behind my interest in philosophy were mixed. Alongside intellectual fascination, I was also looking for ways to justify my trust

in science. I favored those early twentieth-century philosophers who
drew sharp lines between science and enterprises that only pretended to
deliver facts.[26] Grand systems of metaphysics sounded like fake profun-
dities made up by pompous people who tripped over their own words.
Religion was too tangled up in fairy tales about gods and demons. And
the paranormal—I had had a phase of half-belief in UFOs and psychics
and the Bermuda Triangle in my early teens, but I had gotten over it and
thought everyone else should as well. Therefore, I was very interested
in efforts to draw a line between science and nonscience, only to find
that more recent philosophy was getting bogged down or abandoning
the quest. I ran into objective Bayesian statistics through some of my
reading in statistical physics and became enthusiastic—not just because
of its affinity to physics and my mathematical habits of thinking, but also
because it promised a compelling set of foundations for science.[27] I trusted science, but I was looking for a grounding, a justification, a sense of
security. If I were a religious believer, I expect I would have been trying
to find a system of apologetics that worked.

I still trust science today. It would be hard for me to function as a
professor of physics otherwise. But I have just about given up on finding
a method that will guarantee scientific progress. One reason is that I have
seen all forms of that quest get derailed by technical obstacles. Objective
Bayesianism, for example, took the difficulties with initial probabilities
for theories seriously. It proposed to solve the problem by precisely defining the state of complete ignorance before any data becomes available, by minimizing a technical measure of information. Sometimes this
procedure works, and I still use a bit of it when I teach statistical physics.
But it works only when we have a background theory like quantum mechanics to precisely define states and probabilities. Otherwise, we cannot
be precise about ignorance, and the paradoxes and conundrums that undermine overly ambitious claims for statistical inference come rushing
back in.[28]

There is more, however, that worries me. I have come to think that
the problem is deeper than recurrent technical obstacles. The very no-

tion of establishing a foundation for science has to go.

Attempting to secure science through a method handed down by philosophy sets up a game with a predictable set of moves. We justify scientific knowledge by the scientific method, but then what justifies the method? An even deeper, even more abstract principle? What justifies that? It may seem reasonable to say that we need to assume a principle such as the uniformity of nature. But then, a creationist may say that merely expresses the uniformitarian prejudice of evolutionists. Why should we not consider a creationist story which includes discontinuities in nature, because the ultimate explanation lies not in mindless nature but in the designs of a supernatural creator? Creationists often say that they do not disagree with evolutionists about any material evidence; they just happen to interpret the evidence through the lens of different foundational assumptions, different faith commitments.[29] We often conceive of reasoning through the metaphor of a building. Good arguments are securely based on what was already in place, while bad arguments are unfounded or baseless. But then, what is the bedrock for the ultimate foundations? What justifies the justification? Do we really want to say, with the creationists, that it all comes down to an act of faith?

We might then try to justify scientific method by using the scientific method. But then, we could end up reasoning in a circle, the way a fundamentalist cites a holy text to support the divine origin of the holy text. Worse, there may be a logic bomb waiting to go off when we apply a principle to itself. Such self-references, at the root of profound results in mathematical logic such as Gödel's incompleteness theorem, have ended many rationalist dreams. For related reasons, there can be no such thing as a general-purpose learning algorithm, guaranteed to produce progress in all possible circumstances. Every fixed learning procedure, every candidate for method, has blind spots and conditions of failure.[30] It might steer away from creationism but endorse Spiritualism, or, most likely, be unable to handle a claim no one has dreamed up yet.

In that case, the justification game is rigged. The only option is not to play the game—to refuse to seek that kind of justification.

No one comes into the world as a blank slate, knowing nothing about how to obtain knowledge. No one faces a situation where they have to set *all* their knowledge aside and have to reconstruct justified knowledge from the ground up. We are always caught in the middle, with imperfect knowledge, trying to improve what we know—sometimes using plausible principles to figure out details, sometimes checking if patterns we see in our world can be made into more general principles. The building metaphor is wrong. Nothing in our knowledge is foundational. We have some very broadly useful ideas, such as laws of nature or statistical methods, and some more particular ones closer to our acts of poking at things, such as facts and data. And what keeps our structure of knowledge together is not that it rests on a more secure base, but the mutual support of items of knowledge that are much weaker standing alone. I like the metaphor proposed by philosopher Susan Haack, who compares knowledge to those parts we have filled out in a crossword puzzle. Clues are ambiguous, and there are plenty of ways to make errors in filling out a crossword. But as our answers interlock and fit together, our confidence that we are getting the solution rightly increases.[31]

The most secure knowledge I know in physics exists as such a strongly connected, mutually reinforcing network. We think, for example, that we know how electromagnetic interactions work. We have powerful, universally applicable equations and innumerable experiments that follow those patterns. We have reliable technologies such as electrical circuits and radio broadcasting, and a solid track record in solving puzzles both in our labs and when gazing at the stars. As in a crossword where we have filled a large number of interlocking words, it is nearly impossible to rip out an isolated bit of knowledge about electromagnetism without massive damage to our overall network of knowledge about physics. Physicists only need to do small repair jobs on electromagnetism, and work mostly on the less secure parts at the outer edges of our network.

Other aspects of physics, such as dark matter, are much less securely locked in place. They are like words we have penciled into a crossword that might answer to one clue, but we have not yet filled in much else around. Such proposed answers have far fewer interlocking connections, and can much more easily be erased and replaced with a better alternative.

Building up a network of reliable knowledge starts with uncertain steps. In physics, we often know we are getting somewhere when theory and experiment start mutually reinforcing one another. We make assumptions, and check if they work or if they have to be scrapped. As we go through cycles of refining explanations and doing better experiments, there is plenty of opportunity for mismatches between theory and data. But if we get lucky, instead of a vicious circle, we end up in a positive feedback loop where experiment and theory correct one another and converge onto a compelling result. Multiple, independent lines of evidence leading to the same conclusion are similarly convincing, for the same reasons: that convergence did not have to happen, and the fact that it did happen strengthens each of the individual strands leading to the conclusion. Falsificationism is correct that science needs reality checks, and that openness to failure is crucial to scientific success. But this does not mean that science is a regime of logical austerity. Instead, we achieve networks of knowledge that depend on mutual support, constantly building and repairing our networks and trying to expand our reach.[32]

After proposing dark matter, astrophysicists immediately started work to see if dark matter could be linked to our other physical knowledge. For their own independent reasons, particle physicists had already introduced the possibility of undiscovered particles that interacted through gravity but not electromagnetism. For example, to make ideas such as string theory work, every particle had to have a supersymmetric partner; these particles, which included dark matter candidates, could have masses too high to have been observed yet. Moreover, the existence of supersymmetric partners promised to help solve a few other problems in physics, not just provide dark matter. This seemed like a good sign, and the most straightforward theories predicted that the required supersymmetric particles would be discovered in our best particle accelerators.

Here, though, is the interesting twist to the tale: none of this guaranteed that dark matter was correct, or that the required particles would be found. In fact, one of the most important recent results in the search for dark matter is that the accelerators gave us nothing. Physicists were not able to achieve mutually supportive lines of evidence or a positive feedback loop. It is as if we tried a possible answer in a crossword, but it failed to interlock with another answer that is more entrenched because of its connections, and so we have to erase it and try something else. Dark matter remains a strong contender, but physicists are less confident about it today than two decades ago.[33]

And that's it. The reasons that most of our knowledge of electromagnetism is secure, and dark matter is not, are exactly the reasons that the physicists give, no more and no less.[34] Physics is not isolated from the rest of our knowledge, since there are methods and instruments that work well in other natural sciences and beyond, and since much of our physical knowledge has consequences that show up in life apart from the physics lab. But these methods are themselves part of our network of knowledge—they are part of our knowledge about good ways to learn about the world. What security they enjoy is due to the same kinds of mutual support that keeps other theories and facts in place. There is no need for a principle such as the uniformity of nature—nature is sometimes sort of uniform and sometimes not, and what we know about uniformity is part of our overall knowledge, not an underlying presupposition. There are no foundations, only results that are more or less supported by the other results of doing science.[35]

Methods Are Results

I don't know if I will ever feel completely comfortable saying that the only real account for the success of science comes from within science, broadly speaking. At some point in middle school, the notion that scientific method is prior to scientific results might have been nailed too deep into my brain.

And yet, that is where I have ended up: insisting that *methods are results*. Our methods of learning about the world do not stand apart from the world and impart a seal of approval to the results—facts, theories— we establish when we follow the method. What happens to be good ways of learning about the world and what are not good ways are themselves facts about the world.[36]

Consider, once again, my Spiritualist student and her séance that summoned Darwin. In mainstream science, we do not accept a séance as a valid method for testing evolution. We do not accept it as valid for anything, really. But it is not too difficult to imagine a different world where, as a matter of fact, properly conducted séances would produce all sorts of reliable information.

There have been believers in psychic powers who, confident in the power of paranormal means of investigating the world, attempted to do science by clairvoyant means. In the early twentieth century, "occult chemistry" used methods akin to spirit mediumship to produce visions of atomic and molecular structure. These efforts came to nothing, producing little but a few books that are interesting historical curiosities.[37] But let us contemplate a science-fiction scenario, an alternate timeline, where occult chemistry succeeded.

After nuclear weapons were discovered, in this timeline, both Soviet agencies and the U.S. Central Intelligence Agency put funding into using psychic mediums to help along nuclear physics, because every small edge on the Cold War foe was valuable. In the late 1940s, psychics operating on both sides reported that the protons and neutrons that made up atomic nuclei were themselves composite particles, made out of three more fundamental particles bound by a very strong force that, unlike electromagnetism, did not weaken with increasing distance. But the psychic visions did not produce a clear path to turn this discovery into weaponry, so this research was sidelined. Funding dried up, and because this research was very highly classified, it did not influence the physics that was done in public. In the 1960s, mainstream physicists came up with the idea that particles called quarks constituted protons and neutrons, and began exploring the strong nuclear force that bound quarks together. And then, some physicists with high-level security clearances remembered the psychic investigations from two decades ago, and found that the psychics had already outlined the fundamentals of quarks and the strong force. In an effort to claim priority, the Soviets announced their earlier work, and the Americans also quickly published.

Since there were independent confirmations from rival research programs, and the psychic visions closely agreed with the later, more conventional work in physics, it seemed clear that there was something to occult nuclear physics after all. Some skeptics, especially among older physicists, refused to accept that the psychics had gotten it right, suspecting some elaborate hoax. But funding started flowing to psychic investigations of particles, and some younger physicists started collaborating with parapsychologists.

The 1970s and 1980s, in this timeline, saw increasing numbers of psychics helping physicists. Further studies established that, just like conducting a séance in a dark room helped reduce distractions for a spirit medium, it was best if the psychic concentrating on visions of physics was housed in a sensory deprivation tank. The most successful investigations came when three psychics worked in tandem, and only visions that coincided in time were accepted. Meanwhile, theorists came up with a modified version of quantum mechanics where the results of experiments were not random, and postulated waves of consciousness that could interact and resonate with the reconceived wave functions of elementary particles.[38] Soon, particle accelerator experiments came to include, as standard equipment, psychic trackers.

All this is just science fiction. But psychic methods of investigation could have worked. They could have become closely integrated into our scientific knowledge, even taking on the role of trusted instruments. If we do not accept such methods as scientific, it is because to the best of our knowledge, our world is such that psychic visions are useless.[39] In contrast, those instruments and methods that we do trust are in place because the scientific community has learned that they are good for learning. For example, we have figured out that double-blind experiments are very valuable when investigating human subjects. The expectations of both the subjects and the experimenters can influence the results, leading to a lot of worthless data. But in physics experiments on simple objects that cannot have any expectations, blinding is much less of a concern. Physicists only rarely do blinded experiments. Our experimental methods build upon our existing knowledge, whether in physics or psychology, and are validated by the same sort of positive feedback and mutual support in a network that characterizes any other reasonably secure result of science.

If all this is right, the séance that summoned Darwin was not scientific, but only because séances are not capable of doing the sorts of things that Spiritualists claim for them. And if my Spiritualist student could show me otherwise, perhaps by getting psychics to reliably predict novel physical results, I would have to change my mind and try to persuade my colleagues to invest in sensory deprivation tanks alongside accelerators.

That still leaves the question of whether Spiritualist methods could have produced true knowledge even without being scientific. Perhaps. In the alternate timeline I just sketched, I described a tight integration developing between physics and psychics. In a different scenario, we could imagine a much more restricted overlap between conventional science and séance-like practices. Spirits of the dead, for example, might remain the almost exclusive territory of occult science, even after natural scientists are convinced that Spiritualists have it right. In other words, it could have made sense to think of conventional science and occult practices as having mostly separate domains, though they both were reliable within their own territory. But that also is far removed from the world we live in.

My Spiritualist student was, I think, both unscientific and wrong. And the reasons for both are very similar.

Institutions Gone Wrong

A séance is both unscientific and untruthful. But I don't want to automatically equate established science with truth. Scientists can go wrong. The established methods of science can send us down rabbit holes.

Within our sciences, we often construct a truncated history, recounting earlier theories and relating how they were superseded by improved knowledge. This narrative of progress is important, since the reasons supporting our current understanding include explanations of how it is an advance over older alternatives. But the fuller history of science is more complicated, including blind alleys, mistakes, and entrenched investments in false hopes.[40] If a conspiracy theorist accuses scientists of having manufactured a pandemic virus in a bioweapons lab, it is not enough to say that the best scientific evidence indicates otherwise. Science is not a set of abstract propositions; it is the knowledge acknowledged and transmitted by institutions such as university science departments and research labs. Why should anyone trust scientific institutions?

There are also questions remaining about claims that challenge mainstream science, such as creationism, psychic powers, or Bigfoot. Certainly the core of any case against such claims is the set of reasons that scientists and investigators give for thinking the claims are wrong. But are there any commonalities to the kinds of weirdness that get labeled fake science, other than rejection by the scientific mainstream?

Sometimes weirdness has common themes. For example, creationists and psychical researchers both object to the impersonal, materialist direction taken by modern science. Instead, they want to affirm personal agency, whether of a cosmic designer or a spirit not reducible to brain activity, as a fundamental principle of explanation. Both are connected to religious movements, and both inspire accusations that the institutions of modern science are structurally biased against anti-materialist views.[41] But such commonalities are limited. Conspiracy theorists also depend on a hidden personal intent driving events, but emphasize politics more than spirituality. Bigfoot enthusiasts need not be interested in personal agency at all.

Examining popular forms of weirdness, we find that some rejected claims share family resemblances due to their dependence on older conceptions of nature that are no longer favored by science. Some communities committed to rejected knowledge may react similarly to exclusion by scientific institutions, adopting a populist suspicion of expertise. And when digging in their heels to protect their belief commitments, defenders of weirdness may drift away from the scientific attitude of being open to learning from nature. They may publicly present themselves as scientific, while behaving otherwise.[42] We can expect the various communities defending weirdness to draw on similar strategies to protect their core beliefs from scientific criticism. But other than such imprecise generalizations, there is no deeper connection between forms of weirdness. There is no grand unified theory of weirdness.[43]

What, then, of the checklists and diagnostic criteria that are supposed to distinguish fake science from the real thing? After all, we still need practical ways to figure out what knowledge claims to trust. Most of the checklists and criteria are probably still sound, as long as we understand them to be rules of thumb, rather than strict requirements. Simpler explanations are more promising, but simplicity is not a judgment we can make independent of our background knowledge.[44] Natural explanations are better than supernatural scenarios, but mostly because modern science has a history of successfully replacing supernatural claims, such as demon possession, with natural alternatives. There is a virtue in testing theories and being open to failure, even if falsificationism is a dead end. Claims that raise alarms on a diagnostic checklist are not necessarily pathological, but they do not deserve our full trust, even if the symptoms turn out to be due to nothing worse than incomplete knowledge.

If we want to dig deeper, though, I think we have to emphasize institutions.[45] Sometimes the weirdness we encounter is fairly superficial, such as a superstition about the number thirteen being bad luck. Explaining such a belief involves the psychology behind our susceptibility to superstitious thinking,[46] some history of ideas, and an interesting cross-cultural comparison with the East Asian aversion to the number four. But superstitions do not strongly challenge science. The more interesting kinds of weirdness propose to rip out and replace large parts of our network of scientific knowledge. Such weird claims are supported and reproduced by institutions that compete with scientific institutions to win public trust. Seeing how institutionalized claims go wrong can tell us a lot about how to get it right.

Consider, once again, creationism and its many variants. Young-earth creationism demands the most extensive disruption to modern science, not just in biology but much of physics as well. Intelligent-design creationism limits the major damage to biology, taking no position on the age of the earth. Similarly, the Islamic creationism that has become popular in countries such as Turkey ignores the age of the earth, but also puts more emphasis on a view of science as infrastructure for technology.[47] While all forms of creationism are a terrible fit with mainstream scientific knowledge, they are not merely assertions of religious doctrine. Young-earthers periodically launch projects to find a better way to bypass radiometric dating than invoking Noah's Flood. During the early years of intelligent design, proponents promised to support their position through new mathematical tools that would reveal the signature of a designer in data. Today, they support lab work that is supposed to advance biology from a design perspective.[48] Some Muslim creationists seek to bolster their rejection of evolution with a more Islamic philosophy of science. None of these efforts can withstand criticism, but creationism still involves genuine intellectual work.

Belief systems such as creationism also consist of networks of mutual support, but their support is not just in terms of explanation and investigation; they critically depend on moral and religious convictions. A whole religious heritage, even personal salvation, is at stake. Modern science has largely separated moral and fact claims, but for many creationists, this separation is just another aspect of the materialist error that has infected modern science. With a network of beliefs that are highly integrated with social morality but very poorly connected to the body of science, most creationist intellectual effort ends up devoted to apologetics. One sign of this is what critics too often label as unfalsifiability: shifting the goalposts as various specific creationist claims become indefensible.

Creationism is mostly concerned with defending belief, not learning about the world. It does not exemplify a scientific attitude.[49]

Young-earth creationism has organic links with churches, and has organizations such as the Institute for Creation Research promoting it. Intelligent design has strong connections to business conservatism, and its leading institution, the Discovery Institute, functions like a think tank. Turkish creationists are supported by pious businessmen, religious brotherhoods, and Islamist political movements. Supporters of creationism do not just come from religiously conservative backgrounds; they are also upwardly mobile and depend on technology in their work. An elaborate fake science such as creationism is attractive to them because they respect and support technology, and they need to harmonize science with their religious conception of the world.

In the United States, though creationism enjoys considerable popular support, creationist institutions operate on the margins of intellectual life. Not only have creationists not found a noticeable foothold in university science departments, they also are not taken seriously in the nonscientific intellectual high culture. There is some rejection of evolution among applied scientists such as engineers and medical doctors, but their opposition has not been organized and focused by creationist institutions.[50]

In Turkey, creationism has become entrenched in public institutions due to support by ruling Islamists. Opposition to evolution is common in the conservative intellectual high culture, and government support has led to a creationist presence in public secondary education and even in some university science departments. The Turkish variety of Islamic creationism reproduces itself not just through religious organizations but also through public education.

In none of these cases are creationist institutions structured to advance knowledge about biology. Far from being independent of religious, political, or commercial pressures, they serve religious and political conservatism. Our best explanations of the persistence and reproduction of creationist views do not refer to a process of learning about nature.

That, I think, is why a label such as "fake science" accurately describes creationism. It is not because the claim of creation is out of bounds for science; it is because of the pathologies of creationist institutions.

Being Scientific about Science

Not all my students defer to science. Some of the best times in the Weird Science classroom are when a student, perhaps fed up with others assuming that science is authoritative, asks if we should be more critical toward science. I can then fade into the background, and let the students learn from each other. When I express skepticism of paranormal claims, I will sometimes get asked whether skeptics are really consistently skeptical, or if we let mainstream science too easily off the hook. It's a fair question.

If it's a good idea to examine the institutions and communities linked to weirdness to see whether and how they fail to promote accurate knowledge, it is also good to ask similar questions about mainstream science. Challengers of established science often accuse the scientific community of being in the grip of a stifling ideological orthodoxy, such as materialism. Maybe they are right. After all, many skeptics and scientists—myself included—have been motivated by secular ideals of progress. If held dogmatically, such ideals can corrupt knowledge no less than religious conservatism. Maybe our institutional procedures such as peer review do a better job of enforcing conformity than screening out mistakes.[51] Maybe our funding structures, which heavily rely on the prospects for military and commercial applications, compromise our independence.[52]

If our sciences are working as promised, our efforts at poking at the world and constructing explanations will lead to imperfect but still genuine knowledge. When we examine our institutions, what we find might support this picture. For example, we can scrutinize a group of biochemists investigating a new compound for its medicinal properties. Sociologists might find that the incentive structures in biochemical research inhibit researchers from reporting results that may be commercially unfavorable to funders. Or they may find that, even with some problems, the biochemists retain enough intellectual independence. Historians of science can put biochemistry into a broader context, and help us see if the internal narrative of progress in chemistry can be sustained or if it is overly self-serving. Philosophers of science can treat the work of the biochemists as a test case for their proposed explanations of how our sciences advance, and if there's a mismatch, modify both their theories and their assessment of biochemistry. If all goes well for the biochemists, the result will be that their institutional arrangements do not predetermine their results but instead enable them to produce new knowledge. Our best explanations for the results the biochemists publish will include substantial contact with the realities of chemistry. And if not, if the institutional features of biochemistry turn out to be uncomfortably close to those of creationism, then it will be time to change some things.

We should have a science of science—which is really what science studies are up to.[53] If our science of science is any good, it will not just affirm our trust in science, but also be capable of calling some of what we do into question, and even proposing ways to improve the processes of science. For those of us who care about genuine knowledge, this is entirely a good thing.

We should welcome skepticism about skepticism. We should adopt a less heroic image of science, acknowledging that institutional failures can happen, and that such failures are opportunities to improve our knowledge. I am now too set in my ways to change my mind much: I think that mainstream science is mostly right, and that paranormal and supernatural claims are fictions. And still, I love having students who push back against science and hope to defend their favorite intellectual territories from the incursions of scientists. They are the ones from whom I learn the most.

Heads Up

There once was an experimental physicist who happily spent her days in her lab, preparing samples, bombarding them with all sorts of exotic particles, and making impressively precise measurements.

She had a habit of going to lunch together with a psychologist friend who would come by the physics lab on the way to the cafeteria. One day, the psychologist stopped by and found the physicist deep in thought, gazing at a lab table with nothing on it but a single coin. The psychologist asked what was happening, and the physicist explained: "I wanted to generate some random numbers, and I flipped a coin. I found, though, that I was getting only heads—I started out with eleven in a row. That seemed strange, so I inspected the coin, and it was nothing but an ordinary penny. So I went on flipping it, and I got eight more heads. By now, I was irritated, and I substituted a nickel for the penny. I got twelve more heads. Really flustered, I wondered if there was something peculiar about the lab, and I went outside. Still heads. I stopped a student walking by and asked her to flip the coin for me. After a few minutes, my total of heads had become fifty-nine in a row. I came back into the lab, switched to a dime, and soon I had seventy-six straight heads. That's about as probable as singling out a particular metal atom in that coin, vaporizing the coin, and then randomly picking an atom from the vapor and ending up with that same atom you singled out."

The physicist waved her hand at her equipment, saying "I've been testing the coins and the surroundings for hours now. There's nothing out of the ordinary. All these are just normal coins, tossed normally. This shouldn't be happening."

The psychologist thought for a few minutes, and responded: "Well, it seems to me that you've ruled out a random coin toss. That's completely improbable. Maybe the common factor here is you, and the attention you paid to the coin. Since you ruled out all ordinary explanations, maybe this is something up the alley of my colleagues who do parapsychology. You were subconsciously hoping that the coin would always end up heads, since that would be a spectacular result, just the sort of thing an experimental physicist like you would want. You must have had a strong psychic influence on the coin tosses." The physicist raised an eyebrow, and the psychologist exclaimed, "Hey, at least it's an explanation!"

The physicist skipped lunch, contemplating the coin. She was reluctant to keep tossing it for fear that it would continue to come up heads. A theoretical physicist from the same department wandered into the lab, asked what was going on, and heard the same story about the coin tosses. "Hmm," the theorist said, "you do seem to have ruled out any sort of known physical explanation for your sequence of heads. I wouldn't let these results bother me, though. After all, if you toss a fair coin seventy-six times, you'll end up with some sequence of heads and tails. Most of those sequences will be a haphazard mixture of heads and tails rather than something that looks special like all heads or all tails. But that doesn't matter: all of those sequences are equally probable. So whatever exact sequence you had ended up with would have been spectacularly improbable, no different than what you've actually observed. You're only spooked because you think there is something special about all heads, but that's you, not the coins."

"So," the experimentalist responded, "now what?" The theorist suggested, "Flip the coins again. All your instruments tell you the coins have an equal probability of coming up heads or tails, so the frequency of heads and tails should eventually get back to what you expect." He then walked out. The experimentalist stayed behind. She wasn't happy. She realized that if she were to get another thirty heads in a row, the theorist could still make exactly the same argument.

After she returned home at night, the experimentalist told her husband the story. Her husband mulled it over, remarking, "You have two predictions now: the psychologist thinks the heads will continue, and your physicist colleague thinks the odds are still fifty-fifty." He liked to visit Las Vegas once in a while and gamble until he lost a few hundred dollars. He drew on his experience and said, "It seems to me that both are wrong. Tails is long overdue. I'd bet on tails if you were to flip the coin again."

The experimentalist never flipped any of the coins again. She had the coins framed and put up on the wall of her lab. And every now and then someone would ask, and then she would tell the story.

2 BREAKING LAWS

The Dybbuk Box

The university where I teach is located in a small town in the middle of nowhere. A few years ago, a local man bought a Dybbuk Box, supposedly housing a Jewish demon, on eBay. That put our town on the paranormal map. The new owner of the Dybbuk Box started to have nightmares and visions, and strange paranormal events allegedly took place. So, as one does, he wrote a book about his experiences.[1] And then the book inspired a Hollywood movie, *The Possession*.

All this meant enough notoriety that an obscure company that produced low-budget paranormal shows for cable TV decided to send a couple of people over to film for an episode of *True Supernatural*. The owner of the Dybbuk Box had apparently buried it to prevent any more negative effects, but he was persuaded to dig it up. Then someone had an idea that they should bring the Dybbuk Box to a conveniently located university and run some scientific tests on it. My interests in weirdness are well known on campus, so I soon got wind of what was going on. The producers were particularly keen on testing the Dybbuk Box for radioactivity, and to have some samples from the wood of the box examined under an electron microscope. I agreed to play along, provided that some students from my Weird Science class would be allowed to watch the proceedings.

Some of my students were very enthusiastic. They knew about the movie and its local connections—all too often, students are better at absorbing pop culture than their coursework. They were particularly pumped up by the prospect that a TV show was going to be recorded in the same building where we had our regular discussions on weirdness. For a couple of my students, this ended up as the highlight of their semester of Weird Science.

I was more cynical in my expectations, and I wasn't disappointed. On the morning of the TV shoot, we brought the freshly excavated Dybbuk Box into the physics teaching labs. The producer wanted some fanfare to the opening of the box, and the owner expressed some trepidation, but all that seemed out of place in a lab setting. At any rate, I found myself waving a Geiger counter over a box with old but unremarkable contents, observing that we were not measuring anything above background radiation. I might have rolled my eyes occasionally; I certainly slipped into physics professor mode and explained what background radiation was. I needn't have bothered. None of my explanations would make the final cut, except for a part where they asked me if scientists would be open to finding something strange happening around the Dybbuk Box. Why not? I also hinted that the tests we were doing were worthless, but I wasn't subtle enough to have a hope of getting that on screen.

During breaks in the shooting, I talked to the owner of the Dybbuk Box and the TV people about what they hoped to accomplish with the testing. After all, even if we had found radioactivity, its meaning would have been ambiguous. Radioactivity, I gathered, could have helped validate the Dybbuk Box, providing concrete evidence that there was something about the box beyond nightmares and unsettling stories. I surmise that for many viewers radioactivity would have been slightly spooky, even without any paranormal associations. But then, testing also risked bringing the Dybbuk Box into the realm of the ordinary, so that a physicist could slap an equation on the phenomenon and a physician could talk about health effects. On the other hand, maybe radioactivity could have been one of the manifestations of an ancient curse or a trapped demon.

The broadcast episode was the tawdry exercise in mystification I expected. Still, I had my fifteen seconds of fame, and even though I was somewhat embarrassed about it all, I did tell friends and relatives when the episode was about to air. My university's public relations professionals must have thought that there was no such thing as bad publicity. And in the end, I was happy because my students had a live example of a paranormal claim to discuss, as well as a closer look at how media representations of weirdness were fashioned.

My students also noticed the tension introduced when bringing science to bear on a paranormal claim. There is a chance of validation. But for a long time now, scientists have acquired a reputation for taking living nature and freezing it, dissecting it, and shoehorning what remains into abstract equations and tables of data.[2] Our end product is the laws of nature, ironclad boundaries beyond which we declare nothing can pass. But a proponent of the paranormal would expect more from the Dybbuk Box than the cheap entertainment of a shiver down the spine. Even the minor-league miracles associated with the Dybbuk Box can introduce a sense of freedom and boundless possibility. Maybe the vitality and spirit in the world overflows the limits of science. Weirdness may hint at something beyond nature and its laws, even as scientists act as old-fashioned schoolmasters rapping reality on the knuckles if it steps out of bounds.

Moving Violations

At some point in history, when the intellectual division of labor was decided, physics got the job of figuring out the basic patterns by which nature operates. Other areas of knowledge have plenty of rules, but their laws are not as rigid as physics. Those of us interested in living things can study biology, learning to think in terms of populations, variation, and all sorts of intricate details associated with the processes of life. Psychologists experiment with behavior and try to find patterns by statistically analyzing their data. They know that what they study must be rooted in neuroscience, but they usually have to be satisfied with explanations that are only loosely anchored on what must be happening in the brain. Physicists build devices and do math.

In physics, if a cat jumps off a wall, we don't care about the cat's genetics or thought processes. The cat becomes a falling mass close to the surface of the earth, with some individuality captured by a drag coefficient that is frustratingly variable as the cat stretches out or curls up during the fall. The fullest possible account might need to keep track of the cat's biology and psychology, but for a physicist, those details are insanely complicated. Physicists want to strip away such details and focus on what is common to all falling masses. We want to express what we find mathematically and capture what is universal.

Physics, therefore, is a science of simple objects. Our most basic particles, such as electrons, are a bundle of sharply defined physical quantities that make up a short list. There are rigid mathematical rules for how these particles interact. We hope to lay the groundwork for the sciences devoted to insanely difficult objects, getting there by slowly adding layer upon layer of complexity.

Biology also has rules. For example, we should not expect to run into a cat with four legs and two wings. Evolution and developmental biology make it almost impossible to find one in the wild. Even so, we might, in principle, have a winged cat. A mad scientist may bioengineer one in the future. And evolution is full of historical accidents: it is not too hard to imagine an Earth-like planet where six limbs rather than four got locked in the early history of vertebrates, and a cat-like predator evolved where two front limbs became wings. If we are looking for laws of nature, the rule that cats don't have wings is not a good candidate.

Not everything in physics fits either. Introductory physics courses have an unfortunate habit of dumping piles and piles of equations on students. I tell my beginning physics students that I don't care if they forget most of the equations they encounter five weeks after their final exam.

But a handful of times each semester, I will point out that the equation we are playing with is truly important—that it tells us something fundamental about how the world works. The equation for a falling object is not one of them. That is only a starting point before we add complexities, and it only works well when close to the surface of a planet with a thin atmosphere.

Laws of physics have to be universal. For example, information cannot be transmitted faster than the speed of light in a vacuum. That has, as far as we know, no exceptions or qualifications. A few statements like this, and their associated equations, are truly fundamental components of the structure of physics. They describe the ironclad boundaries.

In that case, one of the most effective criticisms of a weird claim would be that it violates a law of physics. Psychic phenomena, such as clairvoyance or telepathy, are not supposed to be limited by space and time. A powerful psychic should be able to make a connection with the object to be intuited, even if it is very far away or vastly separated in time. Psychic influences, therefore, are not limited by the speed of light. Physical broadcast signals in three-dimensional space will weaken as the inverse square of distance, or even faster. Clairvoyance, if real, does not seem to have any such limitation. To a physicist, all this suggests that there is something seriously wrong with psychic claims.[3]

And so it goes with many forms of weirdness. An inventor of a free energy device may look for investors, even though a contraption that produces more energy output than input violates energy conservation. It is a scam.[4] Young-earth creationists who want to believe in the miracles of Noah's Flood and Joshua stopping the earth's rotation end up with enormous amounts of excess heat they can't get rid of, violating energy conservation again. These events never happened.[5] UFO proponents say that flying saucers maneuver with absurdly high accelerations that would flatten any aliens on board. They are not real.

Now, if we were certain about the laws of physics, violations of the laws would be powerful reasons to doubt weird claims. But the history of science shows that our knowledge of physics changes. For example, scientists used to speak of the law of the conservation of mass. Mass could not be created or destroyed: if we had a collection of objects and we made sure that there was no mass transferred into or out of the collection, the total mass could not change. When physicists figured out relativity, however, together with the speed of light limit, it turned out that mass was a form of energy. In high-energy events, such as nuclear reactions, mass is not conserved. Instead, the total energy is the quantity that does not change.

In that case, a defender of psychic phenomena may say that our knowledge of physics is incomplete.[6] Perhaps *meaning* transcends space and time, and psychic consciousness operates at this deeper level of resonating meanings. Broadcast signals and physical particles are limited by the speed of light and inverse-square declines in intensity, but psychic phenomena are about mind over matter. They need not obey the laws that restrict matter any more than nuclear reactions have to conserve mass.

Similarly, the smarter sellers of free-energy devices claim that they do not violate energy conservation but tap into a previously unknown reservoir of energy, maybe the zero-point energy of the quantum vacuum.[7] Flying saucers must have antigravity and inertia-canceling technologies, based on physical knowledge much more advanced than ours.[8]

With some other claims, as with the Dybbuk Box, the attraction may be precisely that of breaking the laws of physics. Creationists want an omnipotent supernatural agent intervening in worldly affairs. How better to demonstrate control over nature than stopping the rotation of the earth without boiling the oceans? Psychic phenomena are supposed to be about mind over matter. How better to demonstrate the freedom and priority of spirit than breaking free of the restrictions on matter?

At the least, weird claims raise a question: why do we trust that a proposed law of physics is universally valid? If it is a generalization from experience, we might be entitled to a presumption that the law will hold, the way we might expect that the next crow we see will be black after seeing nothing but 643 black crows. Physicists, however, think that a law of physics is a lot stronger than such a generalization. Violations of laws are supposed to be physically impossible, not merely surprising. Why?

Bring Me a Symmetry

Astronomers routinely observe light from billions of light years away. Since that light has been traveling for billions of years, astronomers also look into the distant past. When explaining observations from so far away and so long ago, astrophysicists rely on the laws of physics constructed in our earthly environment, using instruments developed and tested in local conditions. How can they be so confident that the same physics applies in circumstances so far removed?

The light from far away contains clues about its origins. Each chemical element has a very particular signature of light wavelengths due to the quantum energy levels of its electrons. These signatures appear in starlight. We can assume that they are produced by our familiar chemical elements, through processes we understand on Earth. That is, we can propose this explanation like a crossword entry, and then search for other answers to interlock and support one another. By doing so, astrophysicists can determine the composition and surface temperature of distant stars, and find that the stars are usually similar to our sun. They also find that the wavelength signatures are shifted toward longer wavelengths, with larger shifts as the observed stars become more distant. The best explanation for this shift is an expanding universe, which makes sense in the context of general relativity, our best theory of gravity and the geometry of space and time. The answers keep falling into place and interlocking with each other.[9]

Some of our evidence for the universality of physics is particularly intriguing, as it tells us about the meaning of physical laws. This has to do with Noether's theorem, which links the laws of physics with the symmetries of nature.

I normally do not tell my physics students the names of physicists attached to the equations we discuss. The content of science does not depend on persons; names can obscure the fact that science is a community effort. Physicists would have figured out relativity without Einstein. The names attached to the equations are not even always correct. Historians of science joke about "Stigler's Law," according to which no scientific result is named after who first discovered it. Naturally, Stigler was not the first to make this observation.[10]

I make an exception with the work of Emmy Noether. Historically, the institutions of physics have not welcomed women, and they still often downplay women's contributions. Plus, Noether's theorem is a profound result which I think should be better known. Many of our most fundamental results in physics are expressed as conservation laws, such as conservation of energy. Our basic equations of physics—our descriptions of particles and their interactions—have symmetries, which mean that they stay the same when something else changes. For example, our basic physics is always the same, no matter what the time. The same experiment under identical conditions should produce the same result, whether the experiment is performed now or a billion years ago. This time-symmetry of our basic physics, according to Noether's theorem, is equivalent to stating that energy is conserved.[11] If we observe energy conservation in our labs now, and we do, this is also evidence that our physics was applicable a billion years ago.

Physics has lots of symmetries, and therefore a list of conservation laws. There is, for example, a spatial symmetry in physics. The same experiment should produce the same result regardless of its location, on Earth or a billion light-years away. The corresponding conserved quantity is momentum. The fact that energy and momentum conservation are among our best confirmed local results gives physicists extra confidence that starlight from far away and long ago can be understood.

Imagine a large, flat, frictionless tabletop. In fact, imagine that the tabletop is infinite. The tabletop is analogous to our basic physics: every location on the tabletop looks exactly the same. There is no way to distinguish one spot from another. Now, let's toss some marbles onto the tabletop. Within any local cluster of marbles, there can be very complicated interactions and an intricate dance of movements. The spatial symmetry of the tabletop will ensure that, if a cluster is isolated from the effects of other, more distant marbles, the total momentum of that cluster cannot change. The details of how the marbles interact don't matter.

Now, with the apparent symmetry in place, an experiment might still show that the total momentum of the marbles is not conserved. In that case, one possibility is that we have overlooked something. Perhaps there is a species of transparent marbles that are hard to see, and everything adds up once they are taken into account. Physicists first suspected the existence of a particle called a neutrino when energy and momentum conservation appeared to be violated in some experiments. But then, there can also be experiments that cannot be understood without breaking the symmetry. For example, there may be a hole in the table. In that case, there will be a special location on the tabletop, distinct from all other locations. Marbles that fall into the hole will just disappear, and momentum will no longer be conserved.

Physicists have a presumption in favor of symmetries because they define a condition of least information. On our infinite tabletop, there are no distinguishing features: it is bland, monotonous, boring. There are no differences between any two spots that have to be noted. Each marble on the tabletop will have a distinct local environment, but that is entirely due to the configuration of the other marbles around, not the tabletop. If, however, there is a hole in the tabletop, it will no longer be quite as boring. There will be a distinguishing feature—more information—that needs to be accounted for.

Symmetries provide a context in which a presumption of simplicity, or lack of information, makes sense. If in our local vicinity we observe momentum conservation, this does not ensure that there is no hole far away, but assuming a hole would be gratuitous. Without further information to indicate otherwise, the odds will be that our local tabletop environment will be a representative sample of the whole.

Nevertheless, a presumption is not a guarantee. Symmetries can be broken. As it happens, physicists also know of plenty of broken symmetries and have a good idea what to expect when conservation laws fail. For example, physicists presumed that nature would exhibit left-right symmetry—that physics would remain the same in a world that was the exact mirror image of ours. In that case, a quantity called parity would be conserved. But much to physicists' surprise, it turned out that interactions involving the weak nuclear force do not conserve parity.[12] We live in a universe with lots of matter but not antimatter, so early in the history of our universe, matter-antimatter symmetry must have broken down. In cosmology, the presumption of symmetry has not always fared well when applied to the shape and structure of the universe, rather than to basic physics.[13]

In many cases, we have a very good account of how the symmetry was broken. Highly symmetric, low-information states are often unstable. At high temperatures, the physical state of a collection of objects—what a cluster of marbles is up to—is highly disorganized and, on average, reflects the underlying symmetry. As it cools down, the collection of objects has to transition to one of many less symmetric low-temperature states. Each of these low-temperature states is equally probable, and which one is realized is random. This process, called spontaneous symmetry breaking, undergirds many physical phenomena, from the magnets on our refrigerators to the distinction between electromagnetic and weak nuclear forces. As the early universe expanded and cooled, the symmetry between the forces was broken, randomly introducing some structure and information into the universe.[14]

There is more. Symmetry has become a central concept in today's physics. Relativity, which is the theory that gives us the speed of light limit and gravity, is formulated based on symmetries that keep physics identical for different observers. Our fundamental forces are derived from specialized symmetry arguments. And the same approaches to symmetries and broken symmetries are very successful in the physics behind many of our high-tech devices. There is an extensive, well-tested, intricately connected theoretical structure underlying our modern concept of the laws of physics. They are not mere generalizations.

My students sometimes think that by weirdness, I mean anything that is spectacularly unusual in everyday terms. To them, there is little difference between science-fiction notions such as teleportation, telepathy, or slipping into parallel universes, and bizarre phenomena at the cutting edge of physics, such as superconductivity. Physicists speculate about extra spatial dimensions that are tightly curled up and hidden, or cosmologies postulating multiple universes with differing low-temperature laws of physics, and all of this sounds just as strange. Moreover, some of the wilder notions in physics today can be as hard to put to experimental tests as any paranormal claim.[15] And yet, the difference between paranormal weirdness and exotic possibilities in physics is immense. Speculative ideas in play in today's physics slot very nicely into the existing theoretical structure of physics. Weirdness does not. Speculative proposals are genuine attempts to solve real puzzles in physics. Weirdness is just weird.

When a weird claim violates the laws of physics, the problem is often deeper than just going against, say, a conservation law. Physicists test physical laws all the time and occasionally have to revise them, as part of the continual reconstruction of science. A weird claim is far more implausible when it challenges the whole structure of physics. A hole in our infinite tabletop might be surprising, but if we discover that our marbles fall down a hole, then we just have to accept that and carry on. Something like psychic powers is more akin to a claim that a marble suddenly turns into a flying cat.

Quantum Magic

If confronting the structure of physics is too much of a burden, there still is the option of trying to make change within the system. Quantum physics, then, might provide an opening for weirdness. After all, quantum mechanics already has a quasi-mystical reputation. Strange quantum phenomena include particles that are also waves, an uncertainty principle that puts limits on knowledge, and particles that can tunnel through what should have been insurmountable barriers. In a quantum world, ghosts who walk through walls might not be so out of place.[16]

I regularly teach our quantum mechanics course, and I like to wear t-shirts with quantum themes as I teach. Every year our physics majors design a new t-shirt, often with physics inside jokes related to quantum mechanics. This also means that on my travels, I am very likely to be wearing a shirt with something prominently quantum mechanical on it. This often attracts "cool shirt" comments. Occasionally, one of my shirts has started a conversation. I acknowledge that yes, I am a physicist, and I teach quantum mechanics. And then, if the person I've met is immersed in a New Agey spiritual subculture, they ask me about universal connectedness, psychic phenomena, or meditative practices. I have to disappoint them, explaining that quantum mechanics is just physics. We do calculations, and we bang at things in the lab. What physicists do with quantum mechanics is very useful in figuring out everything from neutron stars to the materials in the high-tech phones we carry around. But we don't traffic in mystical connections or spiritual practices. It's all very good to want to feel more connected to other people and to save the environment, but I don't see what careless use of the word "quantum" contributes to such ideals.

Now, it is certainly true that quantum physics routinely offends our sense of what should happen. Our physical intuitions work reasonably well for a large land-dwelling animal on a small rocky planet with a thin atmosphere. But outside of the environment in which we evolved, with the very large and the very small, the very fast or very massive, the very hot or very cold, our everyday physical intuitions become very misleading. Advanced mathematical theories and sophisticated equipment are the only tools we have to grope around in the dark.[17]

Before quantum mechanics, describing the state of an object like a marble was straightforward. A physicist would list the marble's position, velocity, how rapidly it was spinning, and so forth. Then, the physicist would have to figure out the forces on the marble, which would determine the past and future states of the marble. The math would have been very difficult, and a physicist would have to develop an extensive toolbox of mathematical tricks and approximation procedures. Still, the list of quantities that constituted the state of a marble was easily visualizable, and in the lab, these quantities could be measured directly.

Quantum states are different. A quantum state is still a mathematical representation of all we can know about an object like a marble, but it is no longer a list of measurable quantities. Indeed, a quantum state cannot be visualized in everyday terms, except partly through imperfect metaphors involving waves. And since in the lab we still measure quantities such as position, quantum mechanics has an extra layer of translation between states and predictions to check in the lab. Even these predictions come in a nonintuitive form: quantum mechanics does not predict definite outcomes but produces a probability distribution over a range of possibilities. The outcome of any individual measurement is random.

When I teach quantum mechanics, I can demystify some of the bizarre features of quantum physics. Particles are still discrete packets of energy, but their state descriptions, from which we derive probabilities for measurements of particle properties, are similar to waves. And if the students are already somewhat familiar with waves, other quantum phenomena become less opaque. The uncertainty principle is not a mysterious limitation of our ability to know; in fact, it is not even an entirely quantum phenomenon—it is just something that happens with waves.[18] The same is true about quantum tunneling through barriers: we can observe ordinary waves doing similar things. For anything much larger than an atom, though, the probability of tunneling through walls is vanishingly small.

Even using wave metaphors, however, it is not possible to have a gut-level familiarity with quantum physics, the way one can become an expert marble player. Even those of us with extensive experience can easily get lost if someone takes away our equations. Most physicists come to accept that quantum mechanics is just like that. As long as we can do our calculations and the results match what we observe in the lab, everything is fine.[19]

There are still some open questions about quantum mechanics; in particular, the famous measurement problem. Quantum states change without any gain or loss of information. In principle, if we could do the math, we should be able to figure out all past and future states if we know the present. But quantum measurement erases the past history of states; therefore, it does not preserve information. Since measurement is supposed to be a physical interaction describable as a changing quantum state, it looks like there is an inconsistency in the foundations of quantum mechanics.

There is a difference, however, between a problem that keeps physicists working and a problem where we give up and cover everything up with mysticism. Quantum measurement is a limiting, approximate description of the results of interacting with a complex, noisy environment and measuring apparatus. In statistical mechanics, which is closely related to quantum mechanics, we already know a lot about how information loss in our limiting, approximate descriptions comes about. Today, the measurement problem does not loom as large as it did in the early years of quantum theory.[20]

Quantum mechanics, then, is another example of a physical idea that is strange in everyday terms, but not in a way that is related to paranormal weirdness. Quantum physics cannot give us faster than light communication, psychic phenomena not limited by space and time, or universal connectedness—whatever that even means.[21] In pop culture and science fiction, "quantum" has become a word attached to anything that seems strange and magical. Real physics is very different from a cheap plot device.

I get especially annoyed with abuse of quantum physics as part of a sales pitch for alternative medicine. Every other quack seems to promise quantum healing.[22] I try to get my students to recognize misuses of physics; I have even asked exam questions quoting alternative medicine websites and asking my students to find the physics mistakes. Recently, I have started to have my students pull out their phones in class and search for "Quantum University." This is an alternative medicine outfit that resembles a real university about as much as their sales patter resembles actual physics.[23] My students are usually not advanced enough to independently see where all the quantum-healing language goes off the rails.

They do, however, seem to have a good eye for online scams.

Now, teaching at a university means that I very readily slip into speaking with the Voice of Authority. So I should acknowledge that among the books defending quantum mysticism and quantum-paranormal connections on my shelves, a number are by authors with physics degrees. In the early years of quantum mechanics, the physicists struggling to make sense of quantum physics even as they were inventing it made use of all sorts of ideas and analogies, some which were drawn from metaphysical or mystical ways of thinking. Those early speculations leached into pop culture and have become fully established.[24] They continue to have residues even in the culture of physics.

A couple of my quantum mysticism books are by physics professors who argue that quantum physics is congruent with Hindu or Buddhist religious convictions. These books do not include any calculations or analyses of data, and they lean heavily on overstretched metaphors attempting to reconcile quantum mechanics with everyday intuitions, which often leads to suggestions of unobservable but almost magical forms of behavior. They then also torture metaphors derived from Indian religious traditions, mixing everything together in a muddle that they hope readers will see as a glorious harmony.[25] I see it all as a forced marriage from which physics is trying to escape.

Back in graduate school, when I first got interested in the shady fringes of physics, I used to wonder how a small minority of otherwise competent physicists could be attracted to quantum mysticism. I wonder no longer. In the past decades I have also encountered physicists sympathetic to intelligent design creationism; I have even known biologists who were outright young-earth creationists. They had conservative Christian, Muslim, or Jewish beliefs.[26]

If I lived in India rather than the United States, I would encounter a lot more quantum mysticism in my academic environment. With the rise and then dominance of Hindu nationalism, quantum abuse can now be found in university science departments, often together with Indian astrology and forms of alternative and traditional medicine. In parts of

Indian academia close to the ruling Hindu nationalists, there is now an institutionalized fake science, much like how resistance to evolution is not unusual among university scientists in Muslim countries.[27]

When weirdness is institutionalized, it becomes much stronger. I don't expect ever to be short of examples of quantum abuse to examine in any of the courses I teach.

A Quantum Loophole

Quantum mechanical outcomes are random: individual events have no pattern or predictability beyond the probability distributions that physicists calculate. The world is a game of dice. This randomness also goes against the image of physics as being ready with an equation for everything. What happens when the underlying reality is just random?

Physics built on a substrate of randomness still has plenty of predictability. For example, the radioactivity I tested for on the Dybbuk Box is described by an equation that says that the measured radioactivity will decrease by a half after a half-life passes. Each individual radioactive nucleus will behave randomly: during each half-life it will have a 50% probability of breaking up and a 50% probability of surviving unchanged. With a very small population of radioactive nuclei, the statistical fluctuations—deviations from the expected or average behavior—will be very large, comparable to the size of the population. A graph of the radioactivity over time can then look quite different from the average behavior described by the half-life equation. But with larger and larger populations of radioactive nuclei, while the size of the fluctuations also grows, that growth is slower than the pace of the population increase. This means that the fluctuations will become smaller and smaller relative to the average. Once we have a visible lump of radioactive material, we will have a population of a gazillion nuclei, and the probability of noticeable deviations from the average will become vanishingly small. Individual unpredictability means rigid predictability for large populations. We can and do bet our lives on the average behavior represented by the equations of physics.[28]

As far as physicists can tell, quantum events are truly random, unlike the outcome of a die roll in a casino. A model of a casino die roll as random, where each face has a one-sixth probability of coming up, is an approximation. The complex interactions of the die with its environment, and the sensitivity of the die to the exact conditions under which its corners bounce off a surface, means that we have practically no ability to predict the outcome of a die roll. But if a misguided federal agency were to give me a few million dollars, I could buy equipment to precisely measure the initial state of the die as it is launched, the local air currents, and the exact details of the surface on which the die will bounce. I could get a powerful computer to solve the equations of motion of the die to the highest degree of accuracy my money could buy. With all this, I expect I could increase my predictive success to, just to make up a number, one die roll in five rather than six. If I had billions of dollars to play with, I might improve my predictive ability by a few more percentage points.

Quantum mechanical randomness is different. I could have resources comparable to U.S. military budgets, and nothing I could do would buy me even the smallest improvement over the calculated probabilities. Quantum physics gives us true, fundamental randomness, which is not due to the limited resources available to a physicist.[29]

Such true randomness is another obstacle to weirdness. If a psychic is supposed to have some kind of paranormal control over or knowledge about quantum events, this cannot be due to quantum mechanics as physicists understand it. That would violate the randomness of individual events.

For creationists, randomness poses a different problem. The randomness produced by physics is a source of novelty for biology, providing the blind genetic variation that natural selection acts upon. Rejecting the possibility of mindless creativity in nature, creationists prefer supernatural intervention. Any outside intervention in physics, however, would risk breaking the symmetries of nature, violating conservation laws. At the very least, some events would no longer be truly random. This is an arbitrary smash-up of the structure of physics.

Curiously, there is a loophole here. Creationists are rebels; they want to break up the current structure of science, which they think has become corrupted by materialism. But there are many moderate religious thinkers who also want to preserve a sense of intelligent design in nature, but who do not want to violate conservation laws. Here is how this could work. An intelligent designer could intervene very rarely, over a very long stretch of time, guiding evolution by making sure that just the right sort of mutations occur that lead to an intelligent species such as humans. These rare tweaks would be totally lost in the statistical fluctuations. Since the quantum versions of conservation laws specify averages and not individual events, such rare interventions would not lead to any detectable violation of physical laws.[30]

Imagine a game of dice in a casino. Unbeknownst to the gamblers, the croupier can surreptitiously turn a dial to control the outcome of dice rolls. The croupier, however, fixes the results only very rarely, and never in favor of the house. Instead, the house loses big to a select few clients with organized crime connections. The whole scheme is devoted to money laundering, and that is how the house gets its cut. If the croupier's interventions are rare enough, any evidence that the dice rolls were not random will be completely lost in the statistical noise. An army of statisticians poring over the records of the dice rolls will not find anything untoward.

Now, if investigators had videos of the croupier acting strangely and turning a dial, and matched such occasions with records of financial transactions involved with the money laundering, they would be entitled to suspect a conspiracy. If they also discovered the dial and figured out how it works, they would have a strong case. But without such external evidence, independent of the record of dice rolls, any accusations against the house will look like they are completely made up. If the accusations include claims that the dial once existed but was removed and destroyed, that security camera footage was doctored, and that financial records were falsified, the investigators will then be advancing a baseless variety of conspiracy theory. Their accusations will be arbitrary, and nothing in the process of formulating the conspiracy theory will include methods known to give reliable results.

Much the same applies to interventions covered up by quantum randomness. The data from physics and biology cannot support a claim of intelligent design, so the case must depend entirely on other evidence. This invariably comes down to faith in revelation: hidden knowledge somehow available to the chosen. The claim is arbitrary. It is formulated to defend a religious tradition, without using processes known to reliably produce knowledge. In other words, the notion of intelligent interference concealed by quantum randomness is a cosmic conspiracy theory.[31] It is almost certainly mistaken.

I appreciate any desire not to violate conservation laws, and I am certainly grateful to moderate religious support for proper science education. And the fact that more respectable theological liberals rather than fundamentalists try to find loopholes for hidden interventions gives me pause. Still, such claims are too close to conspiracy theories for me to give them any credence.

Revolution Through Experiment

I've been emphasizing how physics is a tightly linked body of knowledge, and how its structure is a barrier to weirdness. But no amount of physics can rule weirdness out as impossible. If weirdness is real, and if science is working properly, there should be a way to recognize this reality. Proponents of weirdness regularly accuse mainstream science of unfairly excluding their ideas, of being guided by a materialist prejudice, or at least being stuck in a rigid orthodoxy.[32] Acknowledging weirdness may demand large-scale changes in our body of knowledge, but why not? It would not be the first time a revolution in science took place.

I am wary of claims of scientific revolution, perhaps because they remind me of the overpromising that has become routine in grant applications, or the breathless hyping of breakthroughs that have a reader thinking that scientists usher a new paradigm in with every fall season.[33] Large-scale change in science, such as the adoption of quantum mechanics, is rare. And even such occasions are not as revolutionary as some descriptions of scientific change have suggested. New physics, at any scale of change, comes in against a background of continuity. As quantum physics took over, the kind of work physicists did in the lab remained much the same. Older, pre-quantum theories were not junked: they survive as excellent approximations to what happens in less extreme conditions. Most of our engineering is still based on the old physics, and the old physics is still the bulk of what students learn.[34]

When a revolution occurs, it has a lot to do with data that does not comfortably fit into established frameworks. The experiments that produce such data later become a highlight of the abridged history of progress emphasized within a discipline. Today, students get their first taste of quantum physics by confronting famous puzzles in the history of physics with names like the ultraviolet catastrophe or the photoelectric effect. Such experiments provide a clear signal that something is wrong with the pre-quantum approaches to physics. More important, bringing in quantum concepts solves the problems simply and cleanly, while opening up new horizons to investigate. Jury-rigged models based on the old physics might have covered up the problem, at the cost of lots of arbitrary parameters and hidden influences. But with quantum mechanics, independent forms of evidence interlock and fall into place. In the classroom, physics professors select from and compress decades of the history of physics into a week, but the students can then see why quantum physics is so convincing.

Consider a free-energy device. Free energy, actually, would be nowhere near as revolutionary as the advent of quantum mechanics. Physicists would have no great difficulty incorporating a new source of energy into their structure of knowledge. Right now, the zero-point energy of the quantum vacuum looks like a very implausible source.[35] But a clean experimental demonstration of more output energy than input would establish that physicists need to explain something new and unexpected.

A typical free-energy demonstration, however, is nowhere near so clean. There might be an alleged inventor who is chasing after investors, who wants to demonstrate just enough of how the device works to attract capital, but not enough to risk having his secrets stolen. Having all details checked in a proper lab, with the free-energy apparatus being taken apart and examined, would be unacceptable. Eventually, patents will have to be filed, but that process favors big corporations with expensive lawyers, not an inventor in a garage. For now, the inventor just wants enough investment to go further, ensuring that when it is time for patents, all parties involved will profit by many more millions than otherwise.

All this may be understandable, but it also is a perfect set-up for fraud—much like darkened séances for psychic demonstrations. Chances are, the source of extra energy is a cleverly concealed battery.[36]

The deeper problem is that real experimental environments are much messier than idealized textbook descriptions. Fraud is much less often a problem than unforeseen and unforeseeable mistakes and the numerous uncontrollable small ways in which conditions deviate from the ideal.

Even without any scam going on, it can be very difficult to conduct experiments. The cold fusion fiasco of about thirty years ago, which was similar to a free-energy claim arising from within science, is a good illustration. Achieving nuclear fusion in a chemistry lab did not violate conservation laws; it was merely very implausible rather than almost impossible. Cold-fusion claims surfaced among university chemists, and for a brief period, inspired some ambiguous results from other labs and attempts at theoretical explanation. The prospect of highly marketable results undermined incentives for caution, and many chemists and physicists were briefly caught in a muddle. The fusion results were hard to replicate, but that could have been because of the difficulty of getting the delicately balanced experimental conditions exactly right. Soon, however, it became clear that the original experimenters had not done their work properly—again, because doing the required measurements is quite tricky. Many of the ambiguous near-confirmations were probably because some scientists jumped on a bandwagon and expected to get some sort of result. This sort of thing is not unusual in experimental science. But eventually, the bulk of replication attempts and theoretical arguments indicated that something had to have gone wrong. Cold fusion as a mainstream scientific effort collapsed almost as quickly as it first attracted attention.[37]

The story did not, however, end there. A few investigations into cold fusion are still going on, even today. Over the decades, I would occasionally run into news that small, mostly private investment–driven groups still had hopes, and they would from time to time claim some results. None of their efforts made it to the mainstream journals.[38] I figured that with too much money chasing high rates of returns, cold fusion might have seemed a worthwhile financial bet to some, even if its scientific prospects were not good. With even minimal funding, the occasional borderline experimental result that could not be replicated was not surprising. Cold fusion, then, has become even more like free energy, hanging on the fringes of physical science.[39]

The history of the quest for psychic phenomena has some interesting parallels. Physical scientists were involved in investigating psychic claims during the heyday of Spiritualism in the nineteenth century. The more respectable psychical researchers soon agreed that the best quality evidence would be similar to that relied on in physical science rather than something derived from uncontrolled events such as séances.[40]

Today, psychical research has become parapsychology: a small but scientific effort to show that psychic phenomena are genuine anomalies that mainstream science cannot account for. Parapsychology has all the institutional trappings of any scientific discipline, from peer-reviewed journals to academic conferences. Parapsychologists conduct experiments with considerable methodological sophistication, comparable to straight psychology in their rigor. They produce no end of ambiguous results, and can be very critical of pop culture versions of psychic phenomena, but they also regularly publish results that, if confirmed, would be minor miracles.[41]

There is, however, a problem with seeking anomalies without a theoretical framework for what to expect. Success then depends entirely on achieving a clean, unambiguous signal standing out from any background noise. And in the nonideal, dirty environment of real experiments, made even more difficult by insanely complicated human subjects, the background from which an anomaly should deviate is not well-defined. Pulling winged cats out of a hat would be convincing; no amount of minor screw-ups could produce such a strong signal. The body of parapsychological results is a lot more disappointing; reminiscent, indeed, of cold fusion. There are lots of barely noticeable and inconsistent effect sizes, little success at replication, and no consistent and strong signal that rises above the noise.

In other sciences with insanely complex objects of study, a useful method is to set up a control group. Experimenters can give one group a drug and another a placebo, expecting that the average effects of a messy experimental environment will be about the same in both groups, and therefore any difference will be due to the drug. But psychic powers are supposed to be elusive, capricious, and variable. And since parapsychology is a search for experimental anomalies, it has no theory that can define when psychic effects are present or absent. No control group is possible. So parapsychology keeps producing marginal results that do not stand out from the noise to be expected from uncontrollable minor mistakes. Parapsychology seems condemned, like cold fusion, to linger on without convincing the rest of the scientific community.[42]

Some psychologists think that while the case for psychic phenomena is weak, there are enough intriguing results to warrant continued investigation. Others point to a consistent record of failure since the nineteenth century. I agree with the skeptics. Yes, by the standards prevailing in psychology, parapsychology occasionally shows signs of promise. But psychology itself is full of ambiguous, contradictory experimental results, failures of replication, and areas with weak theoretical understanding. Recently, for such reasons, psychologists have had to confront their own replication crisis. Some of these problems are probably unavoidable with insanely complicated subjects. But whether, for example, a psychological process such as social priming works as previously advertised does not matter for physical laws. It does not have cascading effects through all the sciences. Parapsychology aims for a scientific revolution that would change everything all the way down to basic physics. We cannot judge it according to the looser standards afforded to psychology.[43]

Beyond Physics

At this point, a defender of weirdness may think that if not all science, then certainly physics is rigged against them. In that case, why not try something truly radical? Perhaps weird phenomena are not the sorts of thing that can be captured by physical explanations. If, as with the Dybbuk Box, we are looking for a promise of freedom and boundless possibility, we should then go beyond physics—not just our current understanding of physics, but all possible physics.

But what could lie beyond all physics? Paranormal phenomena are a possibility, but it is hard to be sure. Some forms of the paranormal, like ghosts or the demon that inhabited the Dybbuk Box, involve supernatural agents that may be beyond any reasonable conception of a physical entity. Other paranormal claims, like those investigated by laboratory parapsychology, seem easier to assimilate into some kind of physics— not quantum mechanics, but some new physics of the future perhaps. And in popular culture and science fiction, it all blends together. I am a *Doctor Who* fanatic: not only do I have DVDs of every episode since 1963, I also have the fan network reconstructions of destroyed episodes. As might be expected from a series centered on time travel, the universe of *Doctor Who* is an incoherent mess. But in one episode, the third Doctor exclaims that there is a scientific explanation for everything—including the occult powers of seemingly supernatural creatures, which turn out to be aliens manipulating psychic phenomena. Indeed, supernatural, horror, and science fiction all shade into one another. Certainly in the literary imagination, a sharp boundary between the physical and the paranormal is hard to find. Some things are otherworldly, but the other worlds may have a physics of their own.

Still, the question of the limits of physics is interesting. Personally, I am inclined to think that everything real is also physical: I am not religious, I am skeptical of weirdness, and I obviously have a high opinion of the capabilities of physics. The culture of physics is ambitious, encouraging physicists to poke into everything. So a kind of physics chauvinism comes naturally to me, extending to the modern variety of materialism called physicalism.[44] The problem, then, is that with my prejudices, it may be hard for me to perceive any limits of physics, let alone see what might lie beyond.

Fortunately, there are people far more motivated to limit physics. The intelligent design movement has long tried to establish design as a separate principle that is not reducible to the mindless mechanisms and randomness that characterizes physics. Weirdness can be useful for science: it can raise good questions, stretch scientists' imaginations, and provide challenges that help us refine our knowledge. Therefore, maybe I can let the intelligent design movement do part of my job for me.

Over the last dozen years or so, intelligent design has degenerated into focusing almost exclusively on alleged inadequacies of evolution. The earlier form of intelligent design, however, was more intellectually interesting. Some of its proponents claimed to have invented a mathematical procedure to detect the presence of design in data. Traditionally, materialists have argued that complex phenomena such as mind and life were built from the bottom up, out of mindless physical processes. And these processes were described in terms of chance and necessity— in more modern terms, rules and randomness. Indeed, whatever we do in physics, our basic tools are rules and randomness. Intelligent design proponents took this tradition head-on, arguing that if chance and then necessity could be eliminated as explanations for data, what was left over had to be designed.[45]

I have some sympathy toward this variety of intelligent design, because it has a real argument rather than just a suite of evasive actions. It doesn't seek loopholes or take refuge in cosmic conspiracy theories.

Nevertheless, the argument fails. Physical explanations combine rules and randomness. Evolution through variation and selection is a classic example. Randomness provides novelty: it is perfect to keep a process from getting stuck in a rut. Combining randomness with selection rules produces creativity and adaptation. Indeed, variation and selection has escaped the bounds of biology, helping explain creativity in other contexts such as neuroscience and artificial intelligence. Intelligent design is not a separate principle beyond physics; instead, the processes of evolution are central to creative intelligence itself.[46]

Still, defining physical processes as combinations of rules and randomness is attractive. After all, that is the basic equipment for physical explanations, and any new physics we cook up will just change the ground rules and shuffle up how they combine with randomness. Intelligence might not be beyond physics, but it still is worthwhile to ask what kinds of tasks cannot be accomplished by chance and necessity.

So let me do part of the intelligent design movement's job for them. As it happens, there are meaningful tasks which, appropriately for a quest for an omnipotent designer, require infinite computational resources. For example, most of us today rely on data compression in our daily lives, if only so that our digital photos don't take up an inordinate amount of disk space. There is no such thing as a perfect recipe for data compression that uses only finite resources. If we found good evidence that a device that performed perfect compression existed, this would be proof of a major miracle. No physical process combining rules and randomness could accomplish this task. We would have to take intelligent design and its claims of a nonphysical mind ordering the universe more seriously.

There are some arguments for such miracles of computation, but nothing so far that is remotely convincing.[47] Such claims inhabit the fringes of science, where they occasionally mingle with more familiar sorts of paranormal beliefs. It is safe to say that as far as our sciences can tell, there is nothing in our world that is beyond physics.

Therefore, I like to argue for what I call "chance and necessity physicalism."[48] The world, including insanely complicated objects and traditionally mysterious processes such as life and minds, is made of physical objects and interactions. We may never have a comprehensive, final theory of physics that captures everything we observe. But all our physics will always combine rules and randomness. Quantum mechanics is an example of just such a framework for doing physics. Therefore, we know just the sort of miraculous computation that would violate chance and necessity physicalism and therefore could not be accounted for by any conceivable new physics. We have good reasons to believe no such thing exists. Therefore, according to the best of our knowledge, everything we encounter is physical.[49]

I should not give the impression that all this is an established consensus. Varieties of physicalism are common among scientists and philosophers. Not just my prejudices but much of my own work draws me toward chance and necessity physicalism. But it isn't hard to find better physicists than me who think that physicalism is absurd.[50] When I read their arguments, I end up thinking I am a better philosopher, but I may be wrong. There are many philosophers who argue against physicalism.[51] I end up thinking that I have a better grasp of what physics can do. Again, I may be wrong. I can, however, point to a consensus about intelligent design: it is a failure. If we ever run into anything beyond physics, it will not be thanks to creationists.

Carried Away

Teaching physics, I often worry about students limiting their horizons to what we know well enough to include in the textbooks. So I sometimes suggest that they visit the campus library and look at some physics and astronomy texts from over a hundred years ago. If they take my advice, they will find a lot that they recognize; the bulk of introductory physics is rooted in the eighteenth and nineteenth centuries. And they will also find some content that looks woefully undeveloped, where the speculative ideas of the past no longer make any sense. Just about anything related to cosmology will qualify.

The most ambitious forms of weirdness challenge the laws of physics and fail when tested against the tightly connected, repeatedly confirmed structure of physical knowledge. But since science is a massive collective puzzle-solving enterprise, any assessment of physics should include some of the open questions and uncertain prospects we face today. If there still are physics students a century or more from now, they might look at some of our more speculative ideas today and wonder what we were thinking of.

My decades in physics have not included any major changes; in fact, we have found progress very hard to come by on some of our most important problems. We should have figured out high-temperature superconductivity, I would have thought. Theorists are still banging their heads against it. The question of combining general relativity and quantum mechanics for a theory of quantum gravity had already been much worked-on when I was a student. It still is one of the holy grails of theoretical physics.

Efforts on string theory and its offshoots, which have been the most promising approach to quantum gravity and a complete unification of the fundamental forces, have been going on for decades. There is very little to show other than mathematical progress. The energies needed to directly probe quantum gravity are far, far beyond anything conceivable with our present technologies. We might have hoped for indirect clues from particle accelerators, but disturbingly, even though our accelerators today are many times more powerful than they used to be, they have not discovered much that was not anticipated by models already in place when I was a student. Physics works best when we can get a positive feedback loop going with experiment and theory correcting one another; that is not feasible in all areas of physics right now.[52]

More has been happening in cosmology. We have better instruments and a lot more data, and cosmology has come up with some real surprises in the last few decades. When I was a student, the term "dark energy" was unknown, and we did not speak of an accelerated expansion of the universe. Today, I point out to my students that roughly 70% of the energy in our universe appears to be dark energy and 25% dark matter. Familiar forms of matter and light are just the leftover bits and pieces. Though we have some clues, we do not really know what dark matter or dark energy are; we cannot even be completely confident that they exist.[53]

In many areas of cosmology, we have models that make sense, but we also worry that they have too many free parameters, so that we may be trading one unknown for another rather than coming up with genuine explanations. Most of our models rely on symmetry assumptions about the universe at large distances, but we also regularly get observational results that suggest these symmetries may be broken. I can't answer many of the obvious questions students have about the big bang, because that would require a proper theory of quantum gravity. In fact, I enjoy teaching cosmology because I talk as much about unsolved problems and speculative solutions as what appears to be established knowledge. All these uncertainties are normal in an unsettled, rapidly changing field. Still, I wonder if students looking at our textbooks a few centuries hence will again point to cosmology as the area where we most got it wrong.

Is all well in physics, then? Steady, incremental improvement without many spectacular advances might indicate that we have a lot of things right. On the other hand, maybe physics is stagnating. I really would like to know what is going on with high-temperature superconductivity and quantum gravity before I die.

Meanwhile, we can still ask some questions about the territory where physics shades into philosophy. All knowledge is local, even if our locality can encompass all the observable universe. So even if we figure out something like quantum gravity, we should still expect physics to be ragged at the edges where we approach practical limits. Maybe we will have to think of our universe as just one spacetime bubble among many, and be limited to making some general statistical statements about universe populations, without any ability to directly test universes other than ours.[54] Maybe, as with string theory, we will have to lean very heavily on measures such as internal mathematical coherence in evaluating theories, since more empirical tests are very remote prospects.[55] And in all such cases, our uncertainties will be very large, so that in those do-mains, we may have to give up hope of physical knowledge as solid as the physics of electronic circuits.

Some questions, in fact, may inherently be uncertain, regardless of the progress of physics. Metaphysicians famously like to ask why there is something rather than nothing. Occasionally a physicist will take the bait, observing that that in physics we routinely deal with states with-out particles, including the quantum vacuum. Maybe what we need is a proper physical definition of "nothing," and maybe with some more advanced physics than ours, we will someday be able to talk about a state of pre-geometric chaos in quantum gravity, from which a universe like ours emerges, starting with a quantum fluctuation.[56] Metaphysicians will not be impressed: why, after all, do we have quantum mechanics? Tra-ditionally, ultimate questions are supposed to be answered by the whim of the gods, and it is impolite to ask why the gods had that particular whim rather than another. Appealing to the gods may be an answer, but it is useless as an explanation. And it is hopeless to look for a substitute ultimate answer dreamt up in a lab rather than a temple; even if cooked up by a physicist, it will be the same kind of nonexplanation.

I think we should learn from our experience with randomness in-stead. After all, a series of random results are just those results we have to list individually, without any deeper pattern or explanation showing up in the data itself. Randomness is where we just have to stop and say that is how it is. We inescapably run up against limits in our investigations, where we just have to accept the way that things are. There may well be a deeper layer of physics behind what we see, another step that brings us closer to nothing rather than something. But we are not entitled to think so without evidence, any more than we should favor a conspiracy theory about how the dice are controlled in a casino.

Things are what they are. There need not be a deeper cause. And yet, we have learned a lot. The uncertainties we face about future physics should not deter us from judgments based on what we do know. Weird claims are probably mistaken, particularly when they appear to violate laws of physics. And whatever the shape of our future physics, it is not likely to include psychic powers or intelligent design.

Know-It-All

There once was a scientist who was convinced that there was something to the paranormal. In an experimental spirit, she joined a Tibetan sect to improve her mental discipline and develop her psychic powers.

After long study and intensive concentration, she started to get her meditation techniques right. She started to be able to see concealed objects and affect the rolls of dice, which gave her an advantage in game nights with friends. She learned how to levitate for short periods. And as she progressed further up the stages of psychic advancement, she had a vision of all her past lives.

Then, one morning, she woke up, and knew she had achieved omniscience. If she spent five minutes clearly formulating a question, and went into a trance while concentrating on the question, she could reach a state of utter and complete clarity, where the answer became present to her. In that state, there was no effort, no calculation, no reasoning: she directly perceived the truth, with no mediation through language or any shadow of a doubt.

She wanted to know about the spiders in her house, and soon, she was absolutely certain where each of them were located. In her advanced benevolence and universal compassion, she left them all alone. She wanted to know the truth behind the 9/11 attacks, and as she concentrated, all the details of the conspiracy became transparent to her. And since scientific habits die hard, she wondered if cold fusion was really possible. In her trance, she knew that it was possible, and saw all the technical details that needed to be exactly right before cold fusion happened. She went down to her lab, set up the experiment, and watched controlled nuclear fusion take place on a tabletop.

She checked the solution to a few more scientific mysteries. Then, just out of curiosity, she wondered about the true identity of Shakespeare. The answer came with the usual feeling of oceanic certitude, but while out of her trance, she began to have some doubts. The answer didn't make much sense. So she also concentrated on the question of where convincing evidence could be found. She then traveled to London, and just as her omniscient meditative state had told her, there was the proof in an old stack of letters forgotten in an archive.

So, yes, it seemed she had access to omniscience. Everything always checked out. But that brief episode of doubt also left its mark. Was her omniscient state actually infallible? If she concentrated on a possible planet a billion light years away, which she had no ordinary means of knowing about, she would receive an answer. With her track record so far, it seemed very likely that the answer would be correct. But very likely is not the same as 100% certain. How could she be absolutely sure? Her feelings of uncompromised certainty in her trance were, after all, just feelings. In principle, they could be misleading. Her deep, immediate connection with truth, like supercharged perception rather than a fallible process of reasoning, was also not a guarantee. However unlikely, it was still logically possible that her ability was not 100% absolute ironclad correct all the time. And any omniscience with even the possibility of being mistaken was no omniscience at all.

The more she thought about it, the more worried she got. She was certain, it seemed, that she was not entitled to absolute certainty about her omniscient capabilities. She could, after all, be deceived by a powerful demon who was exploiting her residual greed for knowledge. She could not be absolutely certain that she was spiritually completely pure.

Well, she thought, maybe she should consult her omniscience about her omniscience. But how would that help? If she concentrated on "can I be absolutely 100% certain of the answers I get when omniscient?" she would get an answer. If the answer was yes, wouldn't that be a falsehood, proving that her omniscience was a fake? And if the answer was no, wouldn't that also prove it fake?

She became despondent. But then, should she trust her certainty that it was logically possible that her omniscience could go wrong? It seemed simple and straightforward: if she knew anything about logic, it was logically possible that her omniscience was fake. But could she be 100% absolute ironclad sure that she had not made a mistake in logic?

She stopped eating and drinking, though that was not a problem, since by then she could survive through breathing alone. In desperation, she concentrated on a new question, "Will I go mad if I keep thinking about omniscience?"

She received an answer.

3 GODS AND DEMONS

Religious Questions

I don't lecture to my Weird Science students. I don't expect them to agree with me, and I don't even tell them what I think until the last week of classes. They're not supposed to get into the usual student habit of trying to please the professor. I help, but they have to figure out what they think themselves.

I also don't determine our topics. We spend our first week negotiating what we want to explore, before everyone starts doing their research and preparing presentations for their classmates. During the semester, I learn a lot from my students, especially since I have too little patience for online videos, the source of much conspiracy thinking and weirdness these days. My class helps me keep up with paranormal pop culture.

Very often, my students' interests veer into religious territory. After all, the most prominent institutionalized forms of weirdness have religious associations. Our traditions proclaim the existence of gods, demons, and spirits. They tell us about karma and reincarnation, or promise heaven and threaten with hell. Knowledge about the supernatural comes from visions, prophecy, and revelation. Many religions point to miraculous feats as evidence of their truth. Even our less organized, more individualist spiritualities are often centered on paranormal claims.[1]

And so the class inevitably gets drawn into science and religion discussions. One of our common topics is the creation-evolution dispute. I have had many creationist students who have done well; some of my computer scientists have been particularly attracted to intelligent design as a way of combining their religious commitments and the emphasis on design in their studies. I have had biology majors from conservative religious backgrounds use Weird Science as an opportunity to wrestle with the conflicts they perceive between their faith and their chosen field. I have also had many students who think that science and religion can effortlessly coexist, as long as both stay in their lanes. I try to get them to understand their classmates' struggles—to have them take the difficulties seriously and not settle for a cheap compatibility.

Lately, a more popular topic has been traditional religious miracles and spirit-beliefs. Laboratory parapsychology is all very fine, but looking at various religions seems more genuine, like examining miracles in their natural habitat. Discussions really come to life with international students from countries such as China or Vietnam, who talk about ghosts and spirits in their part of the world. To them, spirits appear not as a component of what Americans would think of as an organized religion, but as a pervasive everyday belief.[2] I have also had American students examine elements of their own traditions that appear implausibly weird to outsiders, such as the Pentecostal students I had look into speaking in tongues.

In the past decade, with the increasing secularization of the United States, I have been getting a higher number of nonreligious students. They often still entertain magical beliefs. I also get students who act closer to a stereotype of a scientific rationalist. I have them read arguments defending supernatural beliefs and urge them to become more sympathetic critics.

It is most interesting to see how the students interact. They don't want to offend anyone, but they also are curious. On one occasion a student announced that she would be bringing a Ouija board to make a demonstration as part of her presentation. Another student approached me later, asking to be excused from that class. He was concerned that something bad might happen; in any case, his conservative Christian sect prohibited him from being involved with occult practices in any way. He stayed away. The Ouija board was a disappointment.

If, however, religions are full of institutionalized paranormal beliefs, they are also a lot more complicated than just organized weirdness. Many students insist on describing the relationship between science and religion in terms of harmony or conflict. Such frameworks do not always work very well.

One of the main reasons is that what we describe as religion is very diverse and messy. An emphasis on belief and doctrines highlights supernatural claims, which can then appear central to the whole structure of a religion. But very often religions are practices and complex social phenomena that are not very different from secular ideologies. In such a context, supernatural commitments look like they have a more peripheral role.

I work in a building that houses a number of science departments. Some of my colleagues in the building are conventionally religious, including all the supernatural bits. Chemists have a reputation of being more devout than biologists and physicists, and my limited experience seems to agree.[3] A lot of my fellow science faculty are secular, in the sense of being indifferent to religious ways of life. A few of them think of religion as a nuisance. But then, some of my colleagues attend a liberal Protestant church. As far as I can make it, this church primarily serves as a social club; many participants are very enthusiastic about the choral singing they get to do. Their camaraderie is bolstered by opportunities to situate themselves and their families within a moral tradition. The churchgoers rarely interpret their faith literally, usually translating it into a middle-class progressive activism. I can't complain. When scientists need help keeping creationism out of public schools, liberal religious people are dependable allies.

To the extent that any religion, no matter how liberal, incorporates supernatural beliefs, I must disagree. My reasons for disagreement are largely those that make me skeptical of any kind of weirdness.[4] But then again, in contexts where supernaturalism does not seem central, it also does not seem worth picking a fight over. Whatever intellectual disagreements may exist between the current state of science and the current varieties of spirituality, I am not sure that igniting a conflict between institutions is a good idea.

Nonetheless, narratives of conflict between science and religion have a strong hold on skeptics of the paranormal as well as on many of my students. Others emphasize harmony, or claim that religion has different concerns than science, such as morality. I think that all of the above are partly correct.

Trust and Commitment

I often run into the claim that science and religion are fundamentally different because they depend on different ways of knowing. Science is supposed to rely on a scientific method, while religion depends on faith.[5]

I am skeptical about a special scientific method. And religious thinkers usually offer reasons for their commitment; even those who emphasize faith tend to deny that it opposes reason.[6] Still, leaps of faith, when they happen, do seem to belong to religion more than science. Supernatural beliefs often depend on intuition, revelation, or mystical illumination. Scientists believe in invisible things such as dark matter or electrons, but that is because of detailed theories and experiments that mutually support each other. The religious believe in invisible spirits or karma and rebirth, but they are likely to offer specially religious experiences or trust in tradition as reasons.

Loyalty to a tradition is not necessarily a bad thing. Anyone who affirms a supernatural belief thereby also declares their trust in a community of fellow believers, elders, and family—people who are trustworthy and who share similar interests and ideals. Most ordinary believers regularly interact with a supernatural realm, by prayer, offerings, and magical healing. But in highly institutionalized religions, specialized religious knowledge is associated with a class of religious experts. Such religions provide pathways to deepen religious knowledge, perhaps to join the experts. There will usually be much study involved. The believer may have to sacrifice many personal ambitions that fall outside of the religious domain. Very often, acquiring religious knowledge requires trust in the sources of religious conviction. If the source of religious knowledge is a teacher, then the teacher demands complete trust. If the source is a set of holy writings, the writings deserve absolute authority.[7] Faith, in this context, is not supposed to be a leap in the dark. Instead, faith is a surrender of self and selfishness, a prerequisite for understanding, and a precondition of rationality. One must believe in order to know, and one must have religious knowledge to earn the trust of the community of believers.

Acquiring religious knowledge is also supposed to be a process of personal transformation. Spiritual knowledge is not an inert collection of facts; it requires progress toward a moral ideal. In alchemy as interpreted by occultists, for example, the quest to turn base metals into gold demanded a purification of the self. Recipes to transform materials were both metaphors for the spiritual development of the alchemist and literal processes that could not produce gold without the accompanying spiritual transformation.[8] Many religious traditions envision similar personal changes, where progress in religious knowledge is inseparable from conforming to religiously defined virtues and burning away worldly vices.

Some traditions focus on sacred texts as a gateway to the supernatural. Many also incorporate mystical practices. Practitioners report transformative experiences that are hard to put into words and that seem like unmediated contact with ultimate realities. The self melts away and a feeling of absolute truth washes over all. Visions may appear, and the visionary may get acquainted with elaborate hierarchies of supernatural beings. The advanced expert acquires miraculous powers; a saint becomes an object of veneration and a source of healing. Even the tombs of dead saints may pulse with magical potency.[9]

Some of this is similar to science. The invisible realities envisioned by religious experts can differ from popular supernatural beliefs, the way a scientific theory is often different from intuitive conceptions of how the world works.[10] Religious experts depend on popular trust, but they serve the rulers by converting political conservatism into moral common sense. Scientists are also an expert class who enjoy public trust, but are funded through commercial and military services to those in power. The institutions of both types of expert are similar, providing an environment where scholarship can enjoy some insulation from outside pressures.

Science can also inspire loyalty, and advancing in scientific knowledge also requires long and arduous training that changes a person. Not everyone can do science; they need not just knowledge but to have internalized a scientific attitude, an ethic of investigation.[11] Scientific work demands a high degree of trust in colleagues. I am not a chemist, but I have had to work with chemists on large projects. Though I cannot check every detail, I am reasonably confident that chemists behave in a scientific manner, don't routinely falsify their data, and remain responsive to evidence. As a result, among its practitioners, science also generates profound feelings of getting it right, of seeing deep below the chaos of everyday appearances.

There are also differences between scientific and religious practices. For example, many religious knowledge claims are far more integrated with morality than scientific claims. Natural science usually hovers around a position of moral neutrality, omitting values from its descriptions of the world. But that difference supports religious claims to knowledge. Many people think that there are moral facts, and that sometimes the rightness or wrongness of a situation is as obvious as the fact that the sky is blue. In that case, it might be plausible that science and religion both produce genuine knowledge in their own domains, and that moral and spiritual facts are the domain of religion.[12] Religious methods of obtaining knowledge are not exactly like that of a chemist, but that is only to be expected. Historians also produce genuine knowledge and have a robust sense of proper evidence and explanation, but history is not usually listed among the sciences. And the tools of a historian include, appropriately, archival investigations rather than laboratory work. If the strength of knowledge has anything to do with richly connected, mutually supportive networks, the way that organized religions closely integrate morality into their knowledge claims has to be a point in their favor.

Religious traditions are also notable for their continual interpretation and reinterpretation of texts and doctrines. The sources of religious knowledge require absolute trust, which risks rigidity. Most sacred and mystical texts are, however, remarkably opaque, contradictory, and open to interpretation. If a reader is convinced that a text is sacred, and the text also does not make sense at face value, this inspires creative reinterpretation rather than rejection of the text. Therefore, religious experts are often evasive in response to criticism, similar to many defenders of weirdness.[13] In contrast, scientists want to make their claims as clear and precise as possible, in order to test and weigh competing explanations. This difference suggests that at least in some instances, science and religion have different purposes. Science pokes and prods and explains, aiming for descriptive accuracy. Religion may be more about achieving meaning. In that case, if they help construct deeply meaningful ideologies, supernatural beliefs may be very useful, regardless of whether they are literally true from a scientific point of view.

Perhaps the most intriguing difference between science and religion is the religious tendency to demand absolute trust. Echoes of this demand can be found in other forms of paranormal belief: psychic powers, for example, are supposed to require immersion and openness rather than a critical attitude. Even the presence of skeptics can allegedly inhibit psychic demonstrations.[14] Absolute trust, however, is very difficult to reconcile with a scientific attitude. Science may not have any pre-established methods set in stone, but everything that has worked so far depends on a critical approach. A religion may claim that supernatural knowledge requires absolute trust and a radical personal transformation, but absolute trust short-circuits the critical weighing of alternative possibilities. Mystics seek to submerge their selves and make unmediated contact with deeper realities. Science distrusts raw experience; it demands that we stand back, not surrender our selves, and ask whether even our deepest convictions can be wrong. Far from obstructing a direct relationship with reality, theories and critical evaluation of theories are crucial for scientific knowledge.

Ways of Unknowing

My students want to know whether supernatural ways of knowing—intuition, mystical illumination, revelation—actually work.

Believers in the paranormal and supernatural tend to prefer intuitive rather than analytic styles of thinking. Many spiritual traditions portray a more critical, analytic approach as a barrier to insight. In some contexts, this may make sense. Artistic creativity, for example, also tends to be associated with intuitive thinking, and also has a weak correlation with paranormal belief. Intuitive thinking is faster, less demanding of resources, and more closely linked to feelings of insight. Under conditions when rapid, low-cost decisions are required, such as when escaping danger, our intuitive responses are best. We settle for a lower cost that comes with a higher error rate. Ponderous, higher-cost analytic thinking works better when we have time and resources and want more accuracy.[15]

Still, even with a higher error rate, our intuitions work well in everyday circumstances. When I put students in an introductory physics lab, I let them rely on their intuitions. They get hands-on experience and witness how physical explanations capture how everyday objects behave. It doesn't matter that our intuitions are not reliable for quantum phenomena. And I don't have them worry about the possibility that their experience is not real but false memories implanted by alien mind-control rays. Everyday intuition and perception is not perfect, but it is a good starting point.

Eventually, scientists have to question the reliability of our senses and intuitions. Indeed, the history of physics has many episodes where physicists have made embarrassing mistakes while trying to detect weak signals. For example, in the early twentieth century, some physicists announced the discovery and then confirmation of N-rays. These rays don't appear in today's physics textbooks, because critics soon found that N-rays were not real. Experimenters were relying on their senses to detect a signal that seemed just barely above the noise that is usual in a messy lab setting. And once the experimenters thought they had a signal, their expectations predisposed them to continue to detect N-rays. Other labs, once they knew what to expect, confirmed the existence of N-rays, much as some labs decades later would seemingly confirm cold fusion. N-rays are now a textbook case of experimenter bias due to prior expectations.[16]

Doing physics properly, then, also means developing a psychological understanding of how to do physics well. We improve our data with better equipment, automation, and by instituting safeguards against mistakes. We try to reduce the role of intuition and perception in performing experiments, even while we welcome creative imagination in designing them. And when we really succeed, we achieve positive feedback loops where we question our methods, improve them, and come back to learn even more about the errors we need to watch out for.

Now, visions, revelations, and mystical episodes are not ordinary occurrences. Many traditions incorporate techniques to induce spiritual experiences, but reserve such practices for an expert class or for special occasions. Spiritual encounters are powerful, often starting a process of personal transformation. I have run into at least one physics student who decided to go to seminary after having a religious experience. And those who have had profoundly meaningful experiences often describe them as imparting certainty: they no longer believe that a certain god exists or that there is life after death, they *know*.[17]

A religious tradition that demands absolute trust may leave things there. A scientist, however, will examine spiritual experiences more critically. After all, any interaction with a reality outside nature breaks symmetries, violating the laws of physics. Feelings of certainty can be mistaken. Hallucinations also feel completely real. Visions and revelations may turn out to be unreliable or useless, just like the occult chemistry of the early twentieth century.

Like hallucinations, supernatural experiences may be due to brains being driven into extreme conditions and unusual states. When ordinary perception is not available to provide corrective feedback, the brain can settle on a model of reality that is very different from what is actually going on. A mystic may feel timeless and at one with the universe because subsystems in the brain that help distinguish self from environment or are involved in time perception are disrupted. Feelings of certainty can be decoupled from their proper function. Even psychedelic drugs can lead to profound experiences, and studying how they affect the brain can help us understand mysticism.[18]

Even without much knowledge of brains, there are reasons to be skeptical of supernatural claims supported by spiritual experiences. A mystic may encounter a Hindu god, or go on a tour of heaven or hell, but it is not possible to independently check such reports. In contrast, if a visionary were to receive the winning lottery numbers with some regularity, that would be impressive. If the god revealed what dark matter is and gave a precise value for the mass of dark matter particles, and this was confirmed, it would be harder to remain skeptical. And if the physical realm is subordinate to supernatural agents, it is a puzzle why such visions very rarely happen and never get it right.

Reports about supernatural realms are not consistent. There are common elements, such as a tunnel of light in near-death experiences, or dissolution of the boundaries of the self in many mystical episodes. These, however, are the elements most easily attributable to similar brains responding to similar unusual conditions. Beyond such commonalities, reports are full of culturally specific details: encounters with gods of the experiencer's own religious tradition, affirmations of particular religious doctrines. Far from being raw, unmediated experiences of ultimate realities, reports of spiritual experiences reveal considerable interpretation and cultural construction.[19]

Religious institutions recognize this problem: while spiritual experience can be a source of vitality, it also needs to be constrained and controlled. Spiritual experiences produce contradicting testimonies about the supernatural. This means a potential for heresy. So organized religions have to strike a balance between an anarchy of incompatible and barely coherent freelance revelations and a stifling orthodoxy. This balance has to be externally imposed: decisions about the authenticity of revelations come from institutions and religious experts. A lot of paranormal belief today is associated with alternative spiritualities, approaching anarchy. Individualist seekers have little sympathy for the rigidities of organized faiths. Scripturalist, puritan religiosity, which often underwrites creationism, is much more disciplined about doctrine. Religious conservatives, such as my student who avoided the Ouija board, distrust alternative spiritualities as much as the materialism that they think infects science.[20]

From a scientific point of view, then, the best prospect for explaining spiritual encounters comes from a combination of unusual brain states and the pervasive influence of culture on human experiences. Now, neuroscience is nowhere near developed enough to provide a recipe for inducing a vision of a Hindu god revealing lottery numbers. Moreover, there are large gaps in our understanding of how brains handle shared cultural representations. Since brains are insanely complicated objects that give scientists headaches, and since there are a variety of experiences that are interpreted as supernatural contact, there is no single, unified neuroscientific explanation for spiritual experiences. Even so, through experimentally produced mystical experiences and studies such as brain scans of meditating subjects, it is possible to draw rough pictures of what happens during unusual experiences. Combining these with what we know about how brains function, we can sketch explanations of what might be going on.[21]

As neuroscience advances, our explanations should improve. Meanwhile, we know enough to say that spiritual experiences are probably not contacts with supernatural realities. A mystic's unshakeable conviction that she has encountered a god is no more reliable than the episodes of euphoric certainty I have driven myself into with mathematical ideas. I still have had to do the calculations, and they have usually failed.

Criticism of spiritual encounters highlights how science does not simply proceed out of raw experience. Our experiences are not separate from interpretations. Even when our perceptions appear to avoid explicit theorizing, the way that our brains structure experience implicitly act as theories. Therefore, it is misleading to describe science as having a set of pure data obtained through empirical means on one hand, and theories that result from processing that data on the other. Theories and data are always intertwined.[22]

As a result, if I want to be scientific, I can never place absolute trust in my experiences. With my cultural background, if my brain was knocked off kilter, I would probably not run into gods. Since I am immersed in science-fiction stories rather than scriptural tales, I might encounter aliens, or even meet monsters from *Doctor Who*. But no matter how real it felt, I would not be entitled to uncritically trust my experience. It might be very difficult for me to overcome my conviction of an otherworldly encounter. But if I wanted to have an accurate description of the world, I would have to try.

The Magic of Trust

Beyond spiritual experiences, there still are the reports of miracles performed by religious experts. This is uncontrolled, wild magic, unconstrained by the stifling environments of laboratory parapsychology. The lack of controls means the evidence is of far poorer quality, but that might be made up for by the sheer spectacular magnitude of supernatural feats. But then, some of the problems with lab parapsychology appear again. Under uncontrolled conditions, the kinds of mistakes, hoaxes, and misunderstandings that can happen in a complex world also grow in magnitude. Once again, we end up with no signal that rises above the background noise.

Skeptics investigating miracle claims uncover many cases of pious fraud, self-deception, and misinterpretations of natural phenomena.[23] But they can't explain all cases. This is common with many forms of weirdness. Many UFO sightings can be explained, and some of the solutions to the more complicated cases read like intricate detective stories. But there is also a residue of hard cases that have not been solved. Many Bigfoot sightings can be explained as anything from misidentified bears to well-planned hoaxes. But not all. Noticing the amount of resources and sheer luck required for cracking some of the more elaborate cases, however, we should expect some unsolved tough cases. Investigators will not always have the good fortune required to have the puzzle pieces fall into place. Similarly, we expect lots of inconclusive murder investigations, and do not take this residue to be evidence for an otherwise hidden cult of human sacrifice.

Therefore, unsolved cases that don't stand out from among the more elaborate solved cases are not impressive. Defenders of weirdness will often argue that at least they have collected enough cases that mainstream science should devote more resources to investigating their claims properly. Often individual scientists will get interested, and a few may even become believers. The collective judgment expressed by the scientific community, however, is that such weird claims are unlikely to reward more intensive investigation.

Without the public evidence that would have been provided by miracles, proponents of supernatural realities may have to double down on demands for absolute trust. Perhaps the methods of science are not appropriate for the spiritual realm. Maybe a critical attitude prevents a personal relationship between a skeptic and supernatural agents. This is an interesting way to evade criticism: build the failure of criticism into the claim itself. After all, no matter what methods our sciences happen to be using, they cannot reveal any truth that includes the failure of those particular methods. There is no guaranteed way of knowing applicable to all possible circumstances.

When confronted by such attempts to protect a belief from criticism, skeptics and scientists often make accusations of unfalsifiability. But while a demand for absolute trust prevents direct investigation of some aspects of a supernatural belief system, there can be plenty of indirect ways to test the belief. For example, my students often ask me what happens to an object that falls into a black hole. Since we cannot recover information from inside a black hole, physicists cannot give a direct experimental answer to such a question. Nonetheless, since our theories always take us beyond the evidence, physicists can still calculate what happens inside a black hole. Evidence that tells us that black holes behave as predicted also gives us reason to trust calculations about their interiors.[24]

A demand of absolute trust immunizes a belief from criticism and enhances its cultural reproduction. But even if absolute trust were an integral part of a supernatural belief system rather than just a protective device, the belief system will still make plenty of claims that make contact with realities that we can investigate. As with claims about black hole interiors, we can still ask if the protected beliefs are part of a system that successfully explains the world. I know of no such examples of supernatural beliefs. Invariably, together with their protective devices, supernatural belief systems look like cosmic conspiracy theories wrapped around unreliable methods of knowing.[25]

Even so, religious demands for trust do present a challenge. For those of us who want our descriptions of the world to be accurate, adopting a consistently critical stance appears to be a good idea. But how much should we want accuracy? Someone who has accuracy as their single overriding purpose is a fanatic, even if they may be an excellent scientist or philosopher. There are other purposes in life, and many involve relationships of trust, demanding that we soften the boundaries of the self and forgo critical distancing. For example, some kinds of political action arise from identifying with others who are oppressed. No supernatural claims need be involved, but the increased vulnerability to error is still present.

Many of us, then, face situations where the demands of love and loyalty take precedence over critical detachment. Most of my students, whatever they learn from Weird Science, remain committed to their spiritual beliefs. They accept a higher risk of error, but in the context of their purposes in life, that risk may be worth taking. And while I am happy with my more critical stance, I have to acknowledge that I pay a price.

Spiritual traditions bring up the question of what sort of person we want to be. This is not really the kind of question science deals with. The answer to such a question is not a matter of fact that calls for descriptive accuracy. Secular philosophies may propose answers as well as spiritual traditions that promote personal transformation. But however anyone resolves such a question, they will have to do more than just carefully evaluate evidence.

Escalating Challenges

Most of science has nothing to say about the supernatural. A polymer chemist will likely never encounter any paranormal or supernatural claims during her daily work. Mystics are not interested in molecular bonds, and the only people claiming a new plastic is a miracle will be advertisers.

Some parts of science, however, are notorious for challenging spiritual beliefs. Evolutionary biology is the most famous example. Conservative Christians, Muslims, and Jews resist evolution. Even many Hindus and Buddhists are uncomfortable with how evolution drains any overall purpose from the history of life forms, preferring a narrative of spiritual progress.[26]

It is tempting to say that creationism is just a misunderstanding. Perhaps fundamentalists can't be helped, but religious moderates do not demand literal interpretations of sacred writings. If a text has six days of creation, moderates can interpret each day to mean long ages. They can read stories such as Adam and Eve as a metaphor, emphasizing the moral message about the kinship of all humanity. At the limits of science, there is still room for supernatural intervention. Perhaps the gods lit the fuse of the universe, and then tweaked evolution to ensure that humans with spiritual capabilities evolved. All the material details, though, are for the scientists to figure out. And for their part, if scientists just stick to the facts and don't attach secular ideologies to their conclusions, they will not trespass on religious territory. Everybody will mind their own business, and everyone but the fundamentalists will be happy.[27]

My students usually favor such a compromise. But neither science nor religion are good at remaining within the limits anyone declares. Evolution overflows the bounds of biology, presenting an escalating series of challenges to supernatural beliefs.

First, evolution contradicts the various creation stories. No revelation ever anticipated geological deep time or the evolutionary relationships between forms of life. At face value, the creation stories are false.

This is a problem because supernatural beliefs are not abstract metaphysical claims. Our religious traditions tell richly layered stories about the purpose and creation of life, the interactions of gods and demons with prophets and kings, and the miracles performed by founder figures. Within such stories, supernatural agents have genuine explanatory roles. Even the more abstract concepts within religious traditions still depend on these background stories to anchor metaphysical intuitions. The gods do not just act in some remote, barely understood sense; their words and deeds are preserved in holy texts. More important, the gods endorse particular ways of life. If we cannot trust the testimony of prophets and founders, our most detailed sources of knowledge about the supernatural order behind the visible world are no longer available. Evolution has been an important reason for the erosion of that trust within our intellectual culture. Our spiritual traditions have made some very important mistakes about the nature of the world.[28]

Moderate reinterpretations avoid blatantly false beliefs. But such reinterpretation gives priority to worldly knowledge. If spiritual sources of knowledge are useless in describing a world that was supposedly created by the gods, this goes a long way toward hollowing out notions of supernatural agents in charge.

Second, humans are related to other life forms. We are not positioned in between mere animals and spiritual beings: we are animals. From a biological point of view, we are not the purpose or end point of evolution but just one of the current forms of life, like cockroaches and cabbages.

One of the most common misconceptions about biological evolution is that evolution is progress toward a preset goal. Monotheistic traditions often portray humans as midway on a great chain of being. The chain starts from dead and passive lumps of matter at the bottom, and then goes onto plants and then animals. Next come humans, then spiritual beings such as angels, and finally the divine.[29] The supernatural claims endorsed by institutionalized spiritualities take on meaning in the context of such hierarchies of existence. But in today's science, such a spiritual hierarchy is useless for explaining anything about biology. There is no chain of being.

Third, according to modern biology, the adaptive features of life forms are explained by Darwinian variation and selection. Genetic variations and the selection process are blind: they do not anticipate any future state; they have no plan or purpose.[30]

Once again, evolution is not progress. Due to explosive diversification and random exploration of different reproductive strategies, we get novelties. For example, while once all life on Earth was as dumb as bacteria, we now have more intelligent life forms, including one species of spectacularly smart ape. That trend toward increasing brainpower is not fundamental to evolution; it is a byproduct of starting from a state that was much simpler. If it turns out that our intelligence is maladaptive in the long term, and we kill ourselves off by nuclear warfare or otherwise trashing our environment, future life on our planet can become much less brainy. Such an outcome is perfectly compatible with our understanding of evolutionary biology.

Religious traditions and derivatives such as intelligent design creationism claim a supernatural purpose behind the history of life. According to evolution, there are no signs of design to be found in biology. By analogy with our own intelligently designed creations such as tools, supernatural forces were once supposed to explain important details about the functional order and complexity seen in life forms. This is no longer possible. Reinterpretations of religion that avoid conflict with science have to retreat not just from traditional claims but also from any claims about biology. They grant nature almost complete autonomy.[31] This is another step toward hollowing out the concept of a supernatural reality underlying the material world.

Fourth, Darwinian variation and selection is not just significant within biology. Physicists deal with simple objects, but they eventually have to ask how adaptation and functional complexity can be assembled out of mindless physical processes. Biologists have already provided most of the answer.

Variation and selection inspires chance-and-necessity physicalism, according to which nature consists of processes combining rules and randomness. Proponents of supernatural design argue that order, life, and mind are imposed top-down, by an intelligence that surpasses all imagination. Our current scientific understanding inverts that picture, so that those features of the universe that appear most interesting to us, such as life and mind, are constructed from the bottom-up by mindless physical processes such as evolution.[32]

If supernatural explanations become useless in natural science, even reinterpreted religious traditions have to depend heavily on spiritual experiences, miracles, and metaphysical assertions for their credibility. Reinterpreting traditions to make them more compatible with natural science is a modernizing impulse. Curiously, the result is to make arguments supporting supernatural realities more closely resemble those arguments available before the development of modern science.

Fifth, an evolutionary perspective, where creativity arises from variation and selection, has begun to influence our scientific understanding of our minds. Not only is our brain a product of evolution, but variation and selection also have important roles in brain functions such as creativity.[33]

Biological evolution has spurred thinking about cultural evolution. We have evolved susceptibilities to perceiving supernatural agents where there are none. These susceptibilities have become a substrate for complex cultural systems and institutions centered on supernatural claims. Today, even our religious traditions, spiritual experiences, and moral perceptions are objects of continually improving scientific explanations in which evolution is an important component.[34]

From ghosts to gods, supernatural agents are supposed to share many psychological features with humans. But when human psychology is increasingly looking like it is built from physical processes, that it has evolved, it becomes harder to imagine a spiritual realm detached from the natural world. This is not to say that that our sciences have complete explanations of what goes on in our heads. As with debates over physicalism, the sciences of the mind cross over into what has been traditionally philosophical territory, and philosophers of mind still have enough controversies to keep them busy for generations. Nonetheless, we have learned enough to further intensify the challenges evolution poses for supernatural claims.

None of this means that evolution and supernatural beliefs are totally incompatible. If I am correct, spiritual traditions do not find any support from natural science. But it is still possible that some other realm of evidence will come to the rescue, or that a philosopher of religion will devise a metaphysical conjuring trick that brings a supernatural realm into existence. Maybe yet another reinterpretation of tradition will save the day.

Most reinterpretations that aim to achieve compatibility with science, however, do little other than suggest how supernatural purposes may be served by allowing the natural world to operate on its own. Knowledge about these purposes is rooted in revelation: hidden powers do things behind the scenes, and a select few are granted knowledge acquired by occult means. This is a cosmic conspiracy theory. Creationists attempt to resolve the conflict by trying to find evidence to overturn evolution. When their efforts are frustrated, they often accuse scientific institutions of denying the evidence because of an ideological commitment to materialism. Scientists must be perpetrating a conspiracy to hide the truth.[35]

Evolution, I think, illustrates how achieving harmony between science and supernatural beliefs is not as easy as everyone minding their own business. Even though scientific and religious institutions are often concerned about very different things, questions about human origins and our place in the universe are of interest to both. And the interpretations that ensure that science and revelation agree about the answers are, in effect, conspiracy theories.

What Do the Experts Think?

My doubts about the supernatural are straightforward. I see supernatural claims as close relatives of paranormal weirdness. Institutionalizing weirdness and embedding it in socially significant traditions certainly adds some interesting twists. But I think that that there is only a short step from skepticism about psychic powers to doubting miracles; a small distance from rejecting intelligent design to dismissing supernatural forces shaping nature.

My position is also controversial. On many forms of weirdness, I can point to a consensus among scientists: creationism is wrong. But this agreement evaporates in the broader public debate. There are also religious experts who endorse creationism, and a few scientists who claim that if done right, science supports creation rather than evolution. And when I argue that evolution presents an escalating series of challenges to supernatural claims, reducing many traditional beliefs to the status of conspiracy theories, I have no consensus to back me up. My point of view is respectable among scientists and philosophers, but even there, it is far from universal. Lots of scientists think that evolution is discomforting only to the more conservative forms of faith. There is no shortage of philosophers who are drawn to high-class metaphysical varieties of supernaturalism, disassociating themselves from low-rent weirdness.

For my students, who often think of science as expert consensus made flesh in textbooks, to be absorbed and then applied in exams, lack of expert agreement can be confusing. If we are discussing UFOs, who are the experts? There are scientists who say that the evidence for alien visitors is not up to scientific standards. Then there are skeptical investigators who pore over popular UFO cases, debunking many by finding mundane explanations. On the side of the believers, my students run into witnesses such as pilots, who swear that what they encountered was completely different from their extensive experience of ordinary phenomena in the sky. There are UFOlogists who often lack academic credentials, but who have spent countless hours examining various cases, and therefore are much more intimate with UFO reports than most scientists. There are even a few scientists who have become convinced that there may be something to UFOs.[36] Whom should a harried student with a paper on UFOs due in a week trust?

Around the middle of the semester, I have a talk with my students, pointing out that a common theme in their work so far is that they are not experts, and they have to negotiate between rival claims to expertise. In fact, figuring out whom to trust is the most important practical question they will confront during Weird Science. A university education, for most of my students, is about acquiring some entry-level expertise, or at least credentials signifying expertise. And no matter what they specialize in, in their professional capacity they will regularly have to make decisions about whom to trust about questions outside of their specialty.

This is certainly true for me. I can claim some expertise on creationist weirdness, but even there, I depend on knowledge about evolution, and I am not a biologist. Because of my interest in creationism, a lifelong fascination with dinosaurs, and a broad curiosity about science in general, I have learned more about evolution than the average physicist. But I am not an expert, and I can't always play it safe and rely on expert consensus. I have, for example, had conversations with a few biologists who were creationists. They knew their field more intimately than I ever could; there were no biological arguments I could offer that they had not heard before. I have also interacted with devoutly religious evolutionary biologists who had thought hard about the challenges evolution presented to their supernatural convictions and thought they could be overcome. I once even had a friendly public debate with such a biologist.[37] In all such cases, I have to combine what I can figure out to the best of my ability with judgments about what expertise I can rely on. My students face a similar situation, but with far less experience and many more conflicting loyalties.

The reasons we rely on expertise are obvious enough. We don't have the resources to check every claim, to analyze every argument. Outsourcing our judgments to communities who specialize in particular types of questions vastly reduces the costs of acquiring knowledge. Not just natural science but every academic and scholarly discipline that claims to produce knowledge is a collective enterprise of research, criticism, and judgment with vastly more resources than any individual. Therefore, expert consensus, when achieved, carries a lot of weight. Experts are fallible, but going against expert judgment without strong reasons is risky.[38]

In that case, why not minimize resource costs and adopt the current expert judgment on everything? My students might prefer this. I already have to fight against their tendency to think that research is a matter of consulting authoritative sources such as textbooks or Wikipedia, summarizing consensus views, and basking in the glory of another completed assignment. If we believe in the wisdom of crowds, we can consult polls to see what the public thinks on creation and evolution. Or, if in a more elitist mood, we can decide that the scientists are obviously the experts, and poll them to see what they think about science and the supernatural.

In the United States, support for creationism is a good proxy for styles of religiosity. In the polls that have been administered for the longest period, the U.S. population has been roughly equally split between conservative religious people who are sympathetic to creationism and more moderate religious people. The moderates agree with the common descent of life forms and with human evolution, but think that evolution is a supernaturally guided process. There is also a smaller number that affirm evolution as biologists understand it: an unguided, natural process. Lately, with the increasing secularity of younger Americans, there has been a growth in support for unguided evolution. Other polls indicate that responses are sensitive to how questions about creation and evolution are phrased. If the questions posed prompt loyalty to a conservative position, creationism looks strong. Otherwise, religious respondents are more open to compromise and moderation; the hard-core support for creationism may not be as high as some polls suggest.[39]

Outside the United States, support for creationism also tracks conservative Christian, Muslim, and Jewish populations. The Muslim example is particularly interesting, though the data about Muslim populations is more limited. Rejection of human evolution is very strong among Muslims, becoming a majority position in large, populous countries that have undergone re-Islamization in the last decades, such as Turkey, Indonesia, and Pakistan. In other countries, the creation-evolution controversy is not as much in the public eye, and opposition to evolution is more muted.[40]

Information from polls is always crude and hard to interpret. I would guess, however, that for most people, the choice between creationism and evolution becomes a question of loyalties. Evolution is an abstract concept that lies buried in science textbooks without having any practical consequences for everyday lives. Very few people have a direct interest in achieving an accurate description of the history of life forms. Therefore opinions on evolution are shaped by trust in conservative religious experts, secular experts, or moderate experts who offer a middle path such as supernaturally guided evolution. Polls about creation and evolution do not reveal scientific knowledge or firm convictions about biology—they probe religious and political allegiances.

When my students in Weird Science discuss supernatural claims with each other, however, merely displaying loyalties does not persuade anyone. They are in a context where they have to offer reasons, and concerns about biological accuracy start to dominate. In that case, scientists might be the proper experts to consult.

According to sociologists of religion, in the United States and Western Europe, scientists tend to be much more secular than the societies in which they live. A majority are indifferent or even hostile to supernatural belief, though some may use the language of spirituality when describing their perceptions of nature. Conservatively religious scientists are a small minority, and they often perceive themselves as operating within a culture that does not welcome their faith.[41]

My experience is much the same. Physics has a very secular culture. I have long been in charge of hosting outside speakers visiting my department, and our dinner conversations often touch on my interests in weirdness. I generally assume my fellow physicists share my perspective, and I am rarely mistaken. I have only been able to find two physicists sympathetic to intelligent design to speak to my Weird Science class, and while they did not quite feel persecuted, they were concerned about reactions from the wider physics community. The small minority of physicists who have reservations about evolution usually come from religious backgrounds.

There are some complications. Scientists in more prestigious positions are often even more secular. And there is a noticeable difference between natural and applied scientists. Notoriously, many medical doctors and engineers are attracted to varieties of creationism. Applied scientists tend to be much closer to the societies they live in, much more tied to business and religious communities. Unsurprisingly, they are politically and religiously more conservative than natural scientists. Applied scientists are also far more numerous than natural scientists. When sociologists of religion want to argue that scientists are friendly toward supernatural beliefs, all they need to do is make sure that lots of applied scientists appear in their sample. Applied scientists even perceive science differently: they more often think of science as a collection of concrete practical facts, downplaying the overarching conceptual frameworks that physicists and biologists emphasize. Frameworks such as evolution become more distant from practice and easier to dismiss.[42]

In India and in Muslim countries, scientists are a lot friendlier toward religion. Such scientists are only slightly more secular than their societies, which are intensely religious. Among scientists, the perception of conflict between science and the supernatural beliefs embedded in religion appears to be a primarily Western phenomenon.[43]

That also fits my experience. Half my family are Turkish, and I grew up in Turkey. My interests in weirdness have extended to Muslim forms of fake science, particularly Islamic creationism. Turkey is not very strong in natural science, emphasizing applied fields; university science departments are often appendages to the education of engineers. During the Islamist political domination of the last few decades, scientific institutions have come under conservative religious influence. In many science departments, especially in newer, provincial universities, a climate of cultural defense prevails, where the harmony of science and traditional Islam is a continual concern. While divinely guided evolution remains an attractive option, creationism is not unusual among science and science education faculty.[44]

What all the surveys indicate, then, is that outsourcing judgments to experts is difficult. There are different, competing claims to expertise. There can be a wide range of views among experts. And much of the variation is due to historical and cultural constraints and institutional features. Relying on expertise is both inescapable and risky.

Surveys of scientists provide some evidence that scientists perceive an intellectual tension between natural science and supernatural beliefs. But directly looking at ongoing debates over supernatural claims would tell us the same. Regional variations, such as the tension rising closer to the surface in Western institutions, are interesting. But scientific institutions in rich and liberal countries are also more successful and have a better claim to independence from nonscientific concerns.

Conducting polls and figuring out a distribution of opinion says little about whether anyone consulted is in a position to know. Prestigious scientists who are leaders in their field cannot also devote a lot of effort to analyzing supernatural claims. So their views need not carry a lot of weight. And worries about cultural influences from outside of science cut both ways. Western science is very secular, but this is not because scientists abandon supernatural beliefs after mastering their science. Instead, already secular people tend to gravitate toward science as a career. I see this self-selection among my students who are physics majors: it is not so much that their science education turns them into skeptics as that many of them are predisposed toward skepticism well before university. That skeptical tendency is part of what draws them to science in the first place.[45] Students with close ties to religious communities and who aspire to serve their community with their talents are more likely to study an applied science with visible practical benefits.

Polling scientists is not a very useful shortcut to answer intellectual questions. The religiosity that remains even in highly secular scientific environments is good evidence for an institutional accommodation between science and religion. It also indicates that the difficulties that theories such as evolution pose for supernatural beliefs are not so strong that they interfere with daily work in the lab. But no more. My students cannot get away with polls and citing experts as a substitute for engaging in argument.

The fact that lots of smart, capable scientists disagree with me about supernatural claims makes me more uncertain. But I cannot outsource my intellectual work either. I screw up, but then so do others. There are many ways in which even scholarly institutions bend to political and cultural pressures. Relying on expertise is supposed to help us reduce the costs of knowledge. But then, evaluating whether claims of expertise can be trusted requires no small amount of work. Studying institutions to see whether they deserve trust is also difficult. New costs start piling up.

We all need to find a balance, which will usually lie between the paranoia of political conspiracy theorists and the absolute trust demanded by some religious experts. If my students learn nothing else from Weird Science, I hope they appreciate that achieving that balance is not easy.

Harmony, Separation, and Conflict

Supernatural claims lead to an intellectual tension between science and religion. This tension is not, however, rooted in any essence or method defining either science or religion. Both science and religion are too fluid, too variable both in geography and in history to have any fixed commitments. Science has not always been distant from supernatural beliefs.

Histories of what became modern science often start with ancient Greece. Greek philosophy asserted its independence from popular piety, and included systematic, critical inquiry into how nature worked. Greek mathematical astronomy was, indeed, impressive, especially in the form it developed after Greek culture came to dominate much of the Mediterranean region. But ancient Greek mathematics and astronomy were closely linked to spiritual understandings of the universe. The Greek philosophers prized mathematics for delivering truths that were certain, independent of the senses. Many of them, especially those adopting Platonic and Neoplatonic philosophies, regularly conceived of human reasoning as a kind of mystical apprehension of timeless and necessary truths. A common motivation for doing astronomy was astrology and astral religions that saw a divine order in the rational perfection of heavenly motions.[46]

In Western Europe, where modern science would emerge, the Christianization and then the disintegration of the Western Roman Empire led to the loss of most of the science and philosophy of antiquity. Christians were almost as destructive in the Eastern Roman Empire, but the damage was not as extensive. Greek Orthodox mystical theology absorbed and transformed much of the mystical philosophy of antiquity. And when Islam arrived in the Eastern Mediterranean, Muslims inherited and further developed the sciences of antiquity. Muslim religious experts were wary about philosophical traditions that harbored rival conceptions of supernatural realms, but supported the "foreign sciences" that promised practical benefits. Some of the sciences of antiquity, such as optics, astronomy, medicine, and mathematics, found their fullest expression under Muslim rule.

Medieval Christian and Muslim science was, however, a premodern form of knowledge where paranormal and metaphysical weirdness of all sorts comfortably mingled with what is, with hindsight, similar to today's science. Astronomy was still tied to astrology, and practices such as alchemy continued to influence investigations of nature. Muslim medical texts from even a few centuries ago are a curious mixture of sound advice on sanitation, useful surgical techniques, and a bewildering array of now mostly dead magical practices linked to both orthodox religion and alternative spiritualities.[47]

Modern science is not just an improved version of its medieval predecessor; it is genuinely new and different. But even at the birth of modern science in Western Europe, ideas about nature combined what we today recognize as science, mainstream religious supernaturalism, and a medley of occult notions. Sciences such as mathematical astronomy were almost mystical activities, attempting to intuit the divine design of the heavens. Isaac Newton spent more of his life on alchemy and biblical prophecy than on physics and mathematics.[48]

Both orthodox and occult supernaturalism continued to find a place in modern science. In the eighteenth and into the early nineteenth centuries, a sense of supernatural design and purpose manifesting in the rational order of the universe, linked to a triumphant Christianity that was colonizing the globe, was explicitly supported by leading scientists. That sense of design came under severe strain in the late nineteenth century, as the conceptual frameworks established by physics and biology matured and came to include Darwinian evolution. Some Europeans began to promote a materialist philosophy inspired by science. But partly as a reaction against such developments, psychical research came on the scene, attempting to use scientific methods to show that there was a human spirit separate from the body that survived death. Psychic phenomena captured the imagination of many scientists and intellectuals.[49]

Keeping paranormal and supernatural notions at arm's length from science is a historically recent development. The overarching conceptual frameworks of modern science, such as evolution and quantum mechanics, leave little room for supernatural agency. The experimental evidence for paranormal phenomena is very weak. But if the paranormal were real, science could have affirmed it. Intellectual traditions emphasizing critical investigation and those based on spiritual illumination could have converged on the same reality. Intelligent design could have hung on; psychical research could have succeeded. Even today, there is a historical residue of the supernatural within science, and attempts to make a case for supernatural realms, though increasingly driven to the fringes of science, will continue. I think it is exceedingly unlikely, but one day we might stumble onto good evidence for realities beyond physics and beyond nature. If that happens, for science, it will be like the return of an old friend—not a stranger or an enemy.

I don't think science has an unchanging essence; I don't think religion does either. For all the claims of ancient and unalterable truths that are embraced by our spiritual traditions, religions can be remarkably adaptable. Many fundamentalist-minded religious movements today are very modern. Their organization, leadership, and advertisements echo corporate practices, their individualism works well with the current economy, and their populist literalism reflects a time when literacy is much more common than in antiquity. There is even an element of empiricism and desire for clarity rather than boundlessly flexible interpretation in fundamentalist beliefs. Fundamentalist stands against ideas like evolution are not just reactions against an increasingly materialist science, they are also products of an intellectual culture deeply influenced by science. And it is not only fundamentalists who are influenced by more scientific ways of thinking. Some of the less organized, paranormal-centered new religious movements of today, whether they are UFO religions or New Agey groups emphasizing psychic powers, have a touch of amateur science about them. Sometimes religions conflict with science because they are close in mentality, not far apart.

Supernatural belief need not be fundamental to spirituality. Though a small minority, secular Jews, Christians, and Buddhists argue that what is most valuable in their respective traditions does not depend on supernatural beliefs.[50] Wisdom traditions centered on meaning and morality do not have to clash with science.

Even in those cases when science as a body of knowledge goes against popular supernatural beliefs, the institutions of science do not need to pick a fight with religious institutions. Large organizations will have reasons for conflict and cooperation, separate from any intellectual tensions that may exist. Modern societies are complex. Our experts in science, religion, and management tend to be independent and based in separate organizations. Many historical instances of friction between science and religion, such as the Galileo affair, are best understood as episodes in the history of scientific institutions establishing their independence.[51]

And yet, no institution is free of entanglements with others. Universities and corporate and government labs are large enterprises that need funding. They serve many other purposes besides figuring out how the world works. Science also draws support from its role in education; maintaining political goodwill about science education is important for the continued health of scientific institutions. Since even in more secular Western countries organized religions have plenty of adherents and political influence, science has institutional motives for seeking to live together in peace. Countering nuisances for science education such as creationism requires help from religious liberals and devout scientists who speak up at public hearings and reassure the audience that science is not a threat. Someone like me, without loyalties to any spiritual tradition, is useless. Pious Muslims usually dismiss my criticisms of Muslim distortions of science, since I am clearly disloyal and I might have a secularist agenda.

Religious organizations also have reasons to avoid conflict. Science is closely associated with the attractions of modern technology. The modern ideal of progress has always had a religious element, and many religious experts find it easier to endorse progress rather than indulge in nostalgia for a time when religion integrated most aspects of culture.

Even in such an environment of shifting institutional interests and opportunities for conflict and cooperation, simplifying myths about science and religion abound. Almost all my students are drawn toward models of the science and religion relationship that have them conflict, have separate spheres, or reinforce one another in mutual harmony.[52]

Harmony between science and traditional supernatural beliefs appeals to religious conservatives. In Muslim countries, this is the institutionally dominant view. Modern technology and economic development are attractive to most Muslims, but the secular Western version of modernity is less compelling. So they try to construct a pious modernity, Islamizing science and limiting it to providing an intellectual infrastructure for business and technology. Desiring cultural authenticity, many Muslims also end up resisting evolution and reviving traditional medical practices such as cupping and using leeches. They are attracted to distortions of the history of science and technology that claim that most of the technological wonders of modern times were appropriated from originally Muslim ideas. In this model, science and religion achieve harmony, but at the cost of science becoming an enterprise of collecting facts like stamps and arranging them to display the majesty of supernatural powers.[53]

Religious liberals and moderates often prefer a model where science and religion occupy separate spheres. The essence of science is its method, which is restricted to providing natural explanations for natural phenomena. Therefore, science has nothing to say about supernatural claims. The peaceful coexistence of scientific and religious institutions follows from the definitions of science and religion. When there is an appearance of conflict, such as that over creation and evolution, the cause must be a misinterpretation of religion or an overextension of science beyond its proper sphere. With such a model, however, the history of debates such as those between Darwinian evolution and its creationist rivals becomes hard to understand. After all, such debates took place within science, and were not settled by philosophical dictates. Intelligent design is a scientific failure, not just a misunderstanding. Separating the spheres of science and religion is useful for keeping the peace, but it is intellectually dubious.[54]

The notion of conflict, even warfare, between science and religion is favored by secular ideologies. Religion, in such a view, is an obstacle to progress, while science is a glorious example of human ingenuity that can perform wonders when not held back by ancient superstition. Historians of science do not find it difficult to debunk such a model.[55] Moreover, an emphasis on conflict is not helpful when trying to limit the influence of fake sciences such as creationism. It turns disagreement about super-natural claims into an all-encompassing antagonism.

When a student asks me whether I think science and religion har-monize, remain separate, or conflict, I'll say all of the above. Then I'll resort to the all-purpose observation among academics: it's more com-plicated than that. And if she's still interested, we will sit down and have a long conversation.

The Simulation

There once was a planet whose inhabitants were devoted to science above all things. As the planetary civilization grew and flourished and then barely survived its polluting industrial phase, it became clear that to limit environmental damage and to prevent future threats, a detailed model of the planet would be necessary.

Therefore, the scientists began a vast enterprise of building a simulation of the planet. The physical scientists started from first principles as they knew them, but simulating every particle would be crazy. Instead, they identified collective behaviors and the emergence of coherent objects, which then could be taken into a higher level of the simulation with their own set of rules. They were gratified to see that out of elementary particles they got to complex biochemistry. Combining biochemical objects with thermodynamics and self-replication, they ended up with individual organisms, then species, and the complex nonlinear interactions that defined societies and ecosystems. They built all these into their simulation. The simulated planet had an atmosphere interacting with the biosphere, the oceans, and the intelligent species whose industrial activity had become a planet-altering force.

The scientists kept adding more feedback loops and more complexity, and more securely anchored higher-level collective behaviors onto simpler lower-level dynamics. They developed new mathematics and advanced their computers beyond all their dreams. And because the best science has plenty of reality tests, they fed vast quantities of data into their simulation. To all the information from all the scientific libraries and databases from their world, they added a continual avalanche of data from the satellites orbiting the planet. The dominant intelligent species had long been used to mass surveillance and their personal data being collected from back when they had cell phones; now, signals from their electronic brain implants were fed to the simulation in order to constrain its operations.

The simulation became so large that its very presence produced significant social and environmental effects. So the scientists invented new mathematical techniques of renormalization to account for the simulation within the simulation. And they figured that to keep runaway feedback effects under control, they should move the simulation off-planet. They built a vast complex for their hybrid ultra-quantum computer arrays on one of their moons.

No matter how good, the simulation remained a crude approximation of the actual planet. Moreover, its mind-boggling complexity and nonlinearity meant that the simulation would always be very sensitive to small perturbations. Therefore, the simulation generated scenarios rather than exact predictions, allowing the scientists to delineate possible futures and calculate statistical likelihoods for the outcomes of policy choices. And this worked fine: there was much rejoicing when governing bodies stopped relying on economic forecasting, and economists were relegated to the status of astrologers.

Nonetheless, after a few centuries of smooth operation and continual improvement, some scientists started to express dissatisfaction. The simulation worked reasonably well, and the projections and statistical likelihoods it generated were trustworthy for decades before the mounting uncertainties made them useless. But the workings of the simulation were now so complex that no one understood why any one scenario ended up more likely than another. The scientists had built a black box that was decent at answering questions about the future, but was not very useful in helping them understand why some futures were more likely.

To help gain some understanding, the scientists decided to incorporate even more information about a small number of simulated individual persons. The life of these representative agents would be tracked in great detail as they went through the simulation. Maybe seeing things from an agent's point of view, as someone within the operations of the simulation, would produce some insight. Many of the scientists became fascinated with the fate of these simulated persons, spending hours every day in front of monitors, tracking the agents' adventures. The obsession of these scientists did not stay confined to the simulation research institutes, and a growing part of the general public also started to follow the simulated agents, now directly hooked up to their visual cortices through their brain implants. Projections from the simulation started to show that a large proportion of the population would continue to do so as their daily entertainment.

Some of the scientists started to worry if this was a good thing. But then, the historians associated with the simulation institutes pointed out that what was happening was not that different from the forms of entertainment and literature common just a few centuries ago, before everyone's talents had to be harnessed to building the planet simulation. The scientists had reinvented fiction. It was mostly harmless.

4 KNOWLEDGE AND MEANING

Between Disciplines

Weird Science is an interdisciplinary course. I get juniors and seniors from all around campus: math and science types, social scientists, humanities people, and business and professional students. They usually start approaching weirdness from their own backgrounds, but they also have to learn something about how to bring in other disciplines. Many questions are not neatly categorized as belonging in one academic box or another, and in such cases, students cannot just rely on the tools they may already have.

Weird Science throws plenty of such questions at the students. So I split the class into groups of two or three, and I make sure I balance the groups. I try to put no more than one math and science type in each group, and also spread out the other clusters of disciplines. Students learn a lot from working with others with different perspectives.

With a mix of students, and questions that don't call for prepackaged forms of expertise, the differences between various academic disciplines is a recurrent subplot. I will have natural science majors raising eyebrows at social science, and business majors wondering how any of what we do may have a practical value. One of my favorite students was an English major who thought that scientists were arrogant: they too readily dismissed the different ways of knowing required to experience paranormal realities. I could count on her to make our classroom discussions livelier. Skeptical criticisms of paranormal and supernatural claims draw heavily on natural science, particularly physics, and on psychology.[1] That might be appropriate. But then again, maybe skeptics have too narrow a perspective.

There is more to the paranormal and fake science than debates about whether any form of weirdness is real. In fact, popular weird beliefs can be an easy target: do we really want to train our big critical guns on Bigfoot? Our best reason to do so might be if Bigfoot is a training exercise. What students learn while debating Bigfoot might help them better analyze other weird claims. And why stop with what is obviously weird? If the students get better acquainted with intellectual mistakes, they might become better equipped to notice intellectual pathologies cloaked in respectability—hiding, perhaps, within their own disciplines. Examining weirdness helps us learn more about the nature of science. We refine our picture of the world, including good and bad methods of learning about the world. And as we do so, we can also turn our critical gaze on less easy targets.[2]

All this requires making use of perspectives from many disciplines. But I admit that sometimes the way we organize knowledge, reflected in the structure of our universities, makes such efforts difficult. I teach at a liberal arts university that takes interdisciplinarity seriously, so much so that I have been able to indulge my interests in weirdness for over two decades while teaching in a physics department. But then, our curriculum also creates the impression that almost every department on campus is responsible for completely separate slices of reality and entirely different methods to investigate them. As faculty, we protect the turf of our own discipline from trespassers, rarely talk to colleagues in different departments, and then complain about how our students struggle to see how our interests and methods overlap.

Our isolation in departments is unfortunate. However we slice up our forms of expertise, we share the same reality. We have different methods: the labwork of a physicist and the archival document-hunting of a historian both need special training, and few will be able to learn to do both well. But since I think that the strength of our knowledge claims depends on the richness of connections in an overall network of knowledge, I also think that it is a good idea to seek further connections. Skeptics, in particular, should draw on more than natural science and psychology. I even like to think of knowledge as a single, somewhat unified network. Physicists will naturally concentrate on experiments and equations, working on our part of the network. We will readily learn from our closer neighbors, such as chemists or mathematicians. But historians and sociologists may also help us understand physics better; they may even offer criticism that can make physicists change some of our ways.

And yet, my English major who thought scientists were arrogant also had a point. The t-shirts I often wear, designed by the physics majors, sometimes poke fun at physics, complaining about the inordinate amount of studying physics demands. But a number of my shirts are not too subtle about proclaiming that physics is superior to other disciplines. Taking doubts about weirdness gained through natural science and applying them on someone else's turf may risk a similar arrogance. Indeed, criticizing other disciplines from a scientific perspective invites accusations of scientism: of assuming that all knowledge can be modeled on natural science.[3] However much I like to wander in between disciplines, I have to acknowledge that I usually have a physicist's mentality, installed by decades of training and practice. I am torn: I don't want to act like an intellectual imperialist, but I also feel like charging full steam ahead.

On balance, I think it is a good idea to apply what we learn from weirdness to other claims to knowledge. Some of my best classroom discussions have taken place when my students have started to make connections between what they have been learning in Weird Science and other intellectual controversies. All our forms of knowledge, I think, have strong continuities with natural science. When any claim to knowledge appeals primarily to intuition or mysticism, it goes off the rails as much as any paranormal fantasy. Attempts to evade criticism or protect cherished beliefs at any cost are rarely good signs. I also think, however, that we should recognize that many of our academic enterprises are not just about seeking specialized knowledge. They may also seek to construct meaning, further a moral ideal, or just propagate an institution. These are all legitimate purposes that can conflict with each other, and even harbor conflicts within themselves. Once again, I find myself thinking that it's more complicated than just figuring out what is the most accurate description of the world.

Meta–Metaphysics

As an undergraduate, I started out as an engineering student. Like many of my peers, I wasn't happy with the requirement that we take a couple of humanities courses—what use would they be for a math and science type? So I cheated. I took a course in mathematical logic, offered by the philosophy department.

I also, however, had a long commute to campus. I started to read philosophy during the ride, starting with Bertrand Russell's *A History of Western Philosophy*. I got hooked. Some of what I read challenged my naïve "science gets reality right" view. I was disturbed, but also stimulated. I wanted proper answers.

My interests in weirdness have led me to reading and doing a lot more philosophy. Much of the philosophy I've encountered has aimed to get our collective conceptual houses in order. We all have intuitions and prejudices about what is and what should be; they sometimes work and sometimes fail. Our intuitions don't always play nice with each other. It would be good to have a disciplined way to reason through and iron out such difficulties.

Philosophy, in this sense, is closely related to the sort of conceptual work that is necessary for constructing theories in science. And the kind of philosophy I like best, which is practiced by the philosophers I have collaborated with, is closely linked to and continuous with the sciences. When we ask about the nature of science or probe what might be wrong with some weirdness, we do something much like a science of science—an explanation of how good explanations work. Going against the stereotype of a philosopher who issues dictates of reason from the armchair, we even have data: the history of science and the various practices of scientists and proponents of weirdness. In many areas, such as the philosophy of physics, biology, or cognitive neuroscience, there is no sharp boundary between philosophy and theoretical science. I find such forms of philosophy attractive because a philosophical perspective allows a more unified, big-picture view, going beyond the more specialized problems in the individual sciences.

Those are the good bits of philosophy. There are also aspects of the philosophical tradition that I find annoying. Within the culture of physics, there is a certain hostility to a caricature of philosophy as navel-gazing, leading to prominent physicists producing a steady stream of comments dismissing philosophy.[4] That is a mistake. But it's also true that philosophy can provide potent excuse-generating tools to protect weirdness from criticism. And if they fall too much in love with reasoning from the armchair, philosophers can produce weirdness themselves.

The trouble often arises from taking intuitions to be constraints on reality. Especially before modern times, a quasi-mystical conception of reason inherited from ancient Greek philosophy caused plenty of mischief. The metaphysicians of old might have suggested that highly refined intuitions about causality, mind, or the gods were direct apprehensions of the structure of reality. Today, the role of intuitions is usually more modest. Intuition is unavoidable and legitimate in the process of getting our concepts in order. In philosophy as in science, we start in the middle of our lives, from whatever half-baked ideas we happen to have. But the potential to get carried away and demand too much of intuition lives on.

Part of the trouble with metaphysics is caused by modern physics, starting with Newtonian physics and gravity. Before the Newtonian era, physics was much closer to everyday intuitions. Every physical theory describes some objects and attempts to explain something about them.

In the physics of premodern times, these objects were usually lumps of matter, interacting through intuitive, mechanical means such as pushes and pulls. In contrast, Newton's formulation of gravity did not provide any mechanism. The theory was very precise and accurate and, therefore, immensely successful. It was universal: it applied to every lump of matter, in the heavens or on Earth. But it was also just an equation, providing a mathematical description but leaving gravity as an action at a distance without assuming any mechanical substrate that would cause gravitational interactions.[5]

Physics kept increasing the gap between everyday physical intuition and theoretical description. Electric and magnetic fields, for example, are not mechanical. They just show up in mathematical models. Moreover, they can't be directly measured. The direction of a field is a convention. We could use different conventions and reverse field directions and get exactly the same experimental predictions. In fact, there are a bunch of symmetry transformations that nontrivially change the mathematical objects in electromagnetic theory, such as fields and potentials, without making any detectable difference. Such symmetries have been essential to our understanding of fundamental forces.[6] And yet, physicists routinely refer to theoretical entities such as fields just like we refer to rocks. Such mathematical objects are inescapably part of our descriptions of the world: they are representations of information about physical states. But representations of information are never unique. Representations can have important differences according to convenience, ease of calculation, or intuitive appeal. These differences matter according to how they might fit the purposes of the community of physicists. Descriptions are always *our* descriptions—they are for our use as well as being about what is out there, independent of what we think.

Quantum mechanics is even more mathematical and far less intuitive. It sets descriptions at a further remove from experimental tests, leaving a large gap between mathematical representations and the realities we can bang our heads against. Even the fundamental mathematical objects can be very different in alternative but experimentally equivalent formulations of quantum mechanics.[7] Once again, the same information can be represented in very different ways. Not everything in theoretical representations can be directly tested. And there can be important pragmatic differences between representations. We might want to directly equate the mathematical objects in a theory with a catalog of objects in a reality that is independent of our choices. But that is very hard to do with physical theories.[8]

This is not just some odd feature of modern physics. Our built-in, intuitive mechanical pictures of everyday objects are also descriptions. Such intuitions seem more perceptual, and do not depend on language or mathematics. And yet, they are not direct apprehensions of reality. Instead, they also are representations, and there are alternative, equivalent representations of the same information. For example, digital music enthusiasts know that a piece of music can also be represented, with no information loss, as a visual map. How we represent information—vision, sound, all our senses—depends on what has been functional in an evolutionary context, for our biological form of life.

There does not seem to be any theory-free, direct access to reality. It's doubtful whether such a notion even makes any sense. And intuitions will have a role in choosing representations that work for us among those that are equivalent in what can be tested.

None of this means that intuitions dictate reality. Consider, for example, the metaphysical intuition that every event must have a cause. This might seem harmless enough, especially since such a principle can be satisfied by attaching undetectable causes to what seem to be uncaused events. For example, we could declare that random quantum events are caused by gremlins looking up results from a predetermined table of random numbers permanently hidden in a dimension Bigfoot inhabits part time. Such invisible causes would be at best immaterial, like the conventional directions of electromagnetic fields, or a distracting nuisance, like the gremlins. And yet, hidden causes would be too easy to confuse with causes that do have a substantial explanatory role. Since prejudices about causality are part of a cluster of intuitions that feed into false claims of intelligent design, they are far from harmless. It is better to recognize that causality in the everyday sense is not fundamental, and that it is built on a substrate of quantum randomness.[9]

The philosophy of religion is perhaps the subdiscipline that is most committed to an older style of metaphysical wrangling. It produces minute variations of very old argument forms, defenses of entrenched metaphysical intuitions in ever so slightly new forms, and a perpetual stalemate. It can be an interesting game to play, especially since arguments over omni-whatever supernatural agents generate paradoxes associated with infinity.[10] But I don't think the philosophy of religion is capable of establishing either the existence or nonexistence of supernatural agents. Moreover, philosophy of religion is curiously disconnected from science and other areas of intellectual life, including the academic study of religion. Not only does it lack strong connections to the rest of our network of knowledge, it does not even encourage interest in establishing any.[11]

I think that reflections on weirdness can help with some constructive criticism of philosophy, reinforcing those self-criticisms philosophers have already explored.[12] At any rate, I am convinced that invoking metaphysical intuition to conclude an argument is the last refuge of a philosophical scoundrel.

Mathemagic

Mathematics, curiously enough, provides an example of metaphysical weirdness that is particularly close to my experience in physics.

There is a common image that associates math with certain and timeless truth, and a magical correspondence to physical reality. Mathematicians and those immersed in math, such as physicists, speak of mathematical concepts like numbers or triangles as if they were physical objects: we refer to them and we speak of their existence. Finding something new about a mathematical object feels like discovering something rather than invention. Historically, the most popular view about the nature of mathematical objects has been some variety of Platonism:

numbers are real things, existing in some sort of timeless realm of abstractions. Since they are not physical, we cannot interact with mathematical objects in any ordinary manner. Our knowledge of math, then, must derive from a special kind of intuition; an occult faculty of direct apprehension.

In theoretical physics, Platonism shows up as the idea that the basic laws of physics, which are mathematical expressions, are the most fundamental reality. The laws are not just descriptions of the patterns in nature; they exist separately from physical objects such as electrons. Math seems uncannily effective in describing physics—maybe this is because reality is fundamentally mathematics.[13]

This is weirdness, though a more respectable kind of weirdness than Bigfoot. Platonic metaphysics highlights the feelings of certainty and discovery associated with doing math, but its account of obtaining mathematical knowledge is akin to supernatural illumination.[14] Instead, we need a proper explanation of how, in physical reality, people work with numbers, and how that develops into a mature mathematical enterprise. What are we doing when we agree that $1 + 1 = 2$? What is a mathematician up to when they produce a proof? Saying that they apprehend a realm of Platonic objects is just mystification. We need an account that draws on our sciences and explains math as a human activity, much like we can understand physics as an activity of probing and explaining.[15]

First, let's deflate some myths. Math is not a way to intuit physical truth. For many centuries, starting with the ancient Greeks, mathematicians thought of Euclidean geometry as the inescapable description of physical space. With the discovery of non-Euclidean geometries, it became apparent that this was not true. Cosmologists today have to consider non-Euclidean, curved spaces. Math may describe the possibilities for physics, but it does not directly produce physical truths.

After that crisis, mathematicians also ran into paradoxes when dealing with infinite sets. They failed to reduce math to logic, and tore their hair out over the axiom of choice. They then found logic bombs in the foundations of math that prevented consistency and completeness in their favorite axiom systems. By the mid-twentieth century, mathematicians who cared about the foundations of their discipline had splintered into multiple rival factions with clashing intuitions about even basic notions, such as acceptable proof. Whatever math is, it is not absolutely certain knowledge established by infallible reasoning.[16]

Still, even if certainty is not available, a lot of math seems pretty secure. And math is not empirical; we do not check theorems about triangles by measuring lots of drawings. But none of this requires the weirdness of magically intuited abstract objects. Consider, for example, fictions. Gandalf, the main wizard in *The Lord of the Rings*, has a beard. That is a true statement, even though Gandalf does not exist. It is very easy to reach agreement about Gandalf's beard: consult Tolkien's novels, or, secondarily, the popular movies. We then discover that Gandalf has a beard, wears a hat, smokes pipeweed, and so forth. Most of us can be far more certain about such statements than scientific claims about physical objects such as electrons.

This is not to say that math is a fiction exactly like Gandalf. The process of creating math is not the same as writing a novel. Math is constructed to be precise, tolerating far less of the ambiguities and shifting meanings in fiction. But then, in either math or fiction, truth can be more indeterminate than we may want. A mathematician may take or leave the axiom of choice, that one can make a new set by taking one element from each of an infinity of other sets. The axiom seems intuitively true, and leads to many important results. But it also has crazy consequences, such as being able to take apart a shape with a finite volume and reassemble the same pieces, without any empty space, and producing a larger volume. In fiction, were Sam and Frodo brought to Osgiliath before entering Mordor? Not in the novel, but it happens in the movie. With the book and the movie being about equally canonical among fans now, it is hard to say there is any fictional truth about the matter. Somewhat like the axiom of choice, it is a matter of choice and convention. In any case, fictions, hallucinations, and mathematical objects are alike enough that a similar task of straightening out intuitions about nonexistent things applies to all.[17]

Mathematics starts with an ability to grasp very small numbers that is present in animals as well as humans. Many kinds of object in our environment come in discrete, countable chunks, like pebbles or chipmunks. Recognizing and making use of this is part of our basic cognitive equipment. Starting with such equipment about number, and spatial and distance relationships, we can then put the metaphor-generating engine in our skulls to work, extending our concepts and generating new mathematical structures. We humans are very good at indefinitely extending our conceptual schemes.[18]

The cognitive basis for math is no more mysterious than our ability to make models not just of observed reality but also of possible futures and alternative pasts and presents, leading to fiction as well as mathematics. But math really takes off when it is recorded, when calculations are performed and shared. Math is a community enterprise, drawing on the abilities of an extremely social and intelligent ape to use language and offer reasons to one another. Math is, ultimately, a social construction.[19]

That may be an odd thing to say: veterans of the Science Wars often associate social constructs with looseness, uncertainty, arbitrariness, dependence on cultural idiosyncrasies. Fictions, explicitly imagined by authors and validated by fans, come from storytelling activities within communities. They are obviously social constructions. But while communities can prize stories that are full of ambiguities and evoke multiple interpretations, they can also aim to reduce ambiguity and reduce the error rate of inferences. Math is full of mechanisms that improve clarity and make inferences reliable. An explicit proof process does exactly that.

Our sciences, especially our so-called exact sciences that heavily depend on math, are similarly constructed. In the theories of physics, we always refer to objects that may or may not be real; in fact, we usually work with approximate descriptions. I can have students determine the direction of a magnetic field in the lab, even though the direction of the field is a convention. They don't directly measure the field, instead observing something more concrete such as the trajectory of an electron beam. And they deploy a pre-quantum concept of fields which is only approximate. In their exam, a question about the direction of a field has a correct answer, much like a question about Gandalf's beard. Math helps us achieve clarity and reliable inferences in science, even if fields inhabit models rather than reality, and even though numbers need not exist in any Platonic sense. We can only grasp the real through the imaginary.

Today's science and philosophy is just starting to create a proper science of culture: of socially shared representations and socially created facts. There is an immense amount of work to be done to connect cognitive neuroscience to communicated and evoked ideas, cultural and biological evolution, and what is already known from the social sciences.[20] Culture is one of those systems that is insanely complicated from the point of view of a physicist. Even a minimally adequate mathematical model of culture would be closer to an immense tangled simulation than an elegant set of differential equations. But I think we know enough to locate mathematics among social constructs such as fictions, games, or money.

There still is the matter of why math is so effective in describing reality. Part of the answer is that it is not effective. Throughout most of its history, math developed together with science, often with the explicit purpose of precisely describing physical reality. Those parts of math are, unsurprisingly, very useful for science. The bulk of pure mathematics, however, has become decoupled from science. Most pure mathematics today is internally driven; describing physical reality is not its purpose, and it is rarely effective at that task.

If, instead, we wonder why the reality we have mapped through our physics is described so well by mathematics, let's turn the question around. What would a world not describable by any possible math look like? A chaos? Something that completely depends on the whims of the gods? An occult universe operating by magic? None of these are beyond mathematics. Math is an indefinitely extensible body, which is not limited to describing relationships of geometric order. Math includes randomness, vagueness, and the fuzzy but resilient inferences of neural networks. Artificial intelligence research constructs ever-improving mathematical models for natural language, so we may plausibly say that English is a subset of math. And as befits infinite sets, math is also a subset of English. In that case, mathematical description is just description—it is unavoidable.[21]

I suspect that wonder at the effectiveness of math comes down to questions about why our world is the way it is, with its simple but often broken symmetries. If so, we can push far with our mathematical physics, but we eventually run into questions very similar to why there is something rather than nothing. And then, we eventually have to accept that explanations come to an end, and the world just is what it is.

Applied Science

Physics is also close to engineering—I may be a refugee from engineering, but I didn't escape very far.

I often have students I advise who intend to transfer to engineering, and I have to explain what they might be getting into. An engineer produces and maintains technology. This means that engineers have to be aware of economic costs and concerns about usability far more than physicists. Engineers can still address deep intellectual questions: the inventors of thermodynamics in the nineteenth century were engineers as much as physicists, and in the twentieth century, information theory arose from solving problems about communication over telephone lines.[22] Still, practical engineering knowledge takes a particular form: to build a bridge across this river, with our resource constraints, engineers tell us what we have to do.

Such concerns entangle applied science with institutional and political purposes. Do the constraints on a bridge include minimizing environmental damage? Who decides to build the bridge, and whom does applied knowledge serve? Applied science always involves purposes other than learning how the world works. These purposes are usually aligned with having accurate knowledge: we need to get our physics right to prevent the bridge from collapsing. But alignment is not guaranteed and is usually not perfect. Achieving the purposes of an applied science requires compromises.[23]

Medicine is a good example of an applied science, not least because it has plenty of associated weirdness. A properly scientific style of medicine is historically recent, and alternative medical practices infused with paranormal beliefs remain very popular. Moreover, medicine becoming more scientific has as much to do with a process of professionalization and attempts to establish an institutional monopoly as with biomedical science demonstrating a more effective understanding of disease.[24]

I live in a town with an osteopathic medical school, and where most medical practitioners are doctors of osteopathy. Their education and their practices are almost all the same as medical doctors, but osteopathy is rooted in a nineteenth-century school of treatment that was as weird as homeopathy or chiropractic. In time, osteopaths decided to move closer to the mainstream of medical science, mainly because it made good business sense. Even today, for marketing purposes, osteopaths sometimes present themselves as a somewhat more holistic, mind-body-spirit alternative to medical doctors.[25]

Some of my premedical students go into osteopathy. And in general, my students face a convoluted landscape of health services. Mainstream medicine is the largest presence in the field, but I will always have a few students who say that they or someone in their family has been helped by chiropractic or naturopathy or alternative medicine of some kind. Alternative medicine is one of our most popular topics in Weird Science.

Invariably, my students are receptive to alternative medical practices. They find the philosophies expressed by alternative practitioners attractive. Attending to the whole person and discouraging heavy-handed interventions with drugs and surgery in favor of boosting natural healing responses seems commonsensical. Mainstream medicine often comes across as the biomedical analogue of car repair, only more distasteful. Proper healing, after all, is more than fixing a bit of wiring or treating a component that has gone wrong.

This does not mean that my students dismiss medical doctors. They generally respect expertise and trust mainstream medicine. They are, however, inclined to seek a compromise. My students are often convinced by arguments that alternative healing modalities should not replace mainstream practices but should be thought of as complementary. They particularly like the notion of integrative medicine. Their research turns up plenty of expert support for such a reconciliation. Famous medical institutions and hospitals accommodate religious and spiritual practices and provide alt-med-lite services for those patients who desire them. The students usually spend some time on the website of the National Center for Complementary and Integrative Health, which is connected to the National Institutes of Health, and discover that some respectable institutions affirm alternative practices.[26]

My students find themselves pulled in multiple directions when they argue about alternative medicine. In doing so, they reflect some of the conflicting interests faced by medical institutions. Healthcare is a business, and in the United States, it is a notoriously predatory business. If there is customer demand for alternative healing, then there is money to be made in satisfying that demand, even at the risk of eroding the scientific reputation of modern medicine. Complementing more invasive treatments with a bit of guided imagery or reiki may even soften the public perception of medicine and distract from the bureaucratic nightmare of health insurance.

The respectability afforded to alternative medicine disturbs many medical doctors and biomedical researchers committed to scientific practices. It bothers many other scientists and skeptics as well.[27] After all, if there is a favorable public perception of science, this has much more to do with practical applications such as medicine than with any low-level public curiosity about dark matter. Both in terms of human well-being and the reputation of science, there is much more at stake with alternative medicine than with Bigfoot. So there has been a movement not just to resist alternative encroachments on mainstream medicine but also to tighten up standards within biomedical science, often under the banner of evidence-based medicine. With randomized double-blind studies, rigorous experimental protocols, and proper statistical procedures, medical researchers should properly vet treatments before recommending them for patients. Healthcare should work, and we determine what works by using the scientific method.[28]

I have no objection to the demand that medicine should be backed by good evidence; I get annoyed when my doctor even hints at anything alternative. But there has been some pushback against evidence-based medicine from within the medical profession. Doing proper experiments is all very well, but evidence-based medicine also can appear as overly rigid and formulaic. Some physicians think that aspects of care for patients and the element of clinical judgment that is vital in complex, uncontrollable environments can be overlooked in all the talk about experimental design and statistical tests.[29] I have some sympathy with such concerns. I think I know something about proper uses of quantification, and a lot of the metrics and measurable benchmarks I see in administrative contexts strike me as devices for bureaucratic ass-covering. I would not like to see more deskilling and the diminishment of professional judgment in medicine. A conception of scientific method as a preset procedure to crank out truth only intensifies my worries.

Some defenders of alternative medicine, however, go so far as to associate demands for standard forms of evidence with fascism. Evidence-based medicine, in their view, is scientism, recommending methods that are inappropriate for evaluating a healing process that is intrinsically spiritual. Alternative medical care is fundamentally about a relationship between a healer and her patient, and that relationship has to be grounded in trust. Procedures such as double-blinding sever that connection: instead of fairly testing a paranormal form of healing, they make such healing impossible. Worse, based on a rigid and illegitimate conception of evidence, critics of alternative medicine try to freeze alternative practitioners out of the marketplace, putting pressure on insurers and inviting state intervention. They want to impose a uniform practice on everyone, substituting a scientistic ideology for freedom of choice.[30]

I am not impressed with such accusations; they are, I think, over-heated versions of evasive behavior very typical of apologists for all kinds of weirdness. It would be a peculiarly fragile act of curing disease that immediately vanishes in properly controlled studies. It is much more likely that measures such as double-blinding make it more difficult for patients and healers to unintentionally enter into a mutually deceptive relationship about the effectiveness of healing attempts.

There is, however, still the complication that health is not a concept that exclusively applies to biological function. Throughout the history of medicine, health has also had social and spiritual connotations. Even today, our encounters with medicine are very often due to ailments that cannot be fixed. Carving out a social place as a suffering person, reconciling one's self to a problem that cannot be fully cured, and finding some meaning in the ordeal are also aspects of medicine. Such aspects are not always captured by applied biomedical science, especially when it acts as if treating disease is analogous to repairing a car.

Alternative medicine is useless in curing cancer, and a menace if, like resistance to vaccination, it undermines public health measures in a pandemic. Most of alternative medicine seems fake, and I am confident that whatever bits and pieces that might work would do even better if assimilated and understood by biomedical science. But curiously, it is harder to criticize alternative healing when it takes on a more religious coloration. It still does not fix biological functions, but sometimes patients adopting the spiritual outlook accompanying an alternative practice can give a more cosmic meaning to their otherwise pointless suffering.[31] Spiritual healing can change the purposes of the patient and redefine success in battling a disease. Life can be chaotic and unfair, and saying that disasters can happen for no satisfying reason is more accurate. But the broader purposes of health and the purpose of being accurate can diverge from one another.

The aims of medicine are not exactly the same as the aims of biomedical science. But the potential for a misalignment of interests is even higher when the healthcare industry acts as a resource-extraction racket.

My students usually try to reconcile alternative and mainstream medicine. But the most reliable way I have found to get them to start asking more skeptical questions is to appeal to their social roles as a consumer. If they have a serious problem, is an alternative practice a good use of their resources? They eventually find out that alternative medicine has become a big business, and some of my students come to think about the holistic rhetoric of alternative medicine as a sales pitch. They become less eager to endorse alternative medical claims. Unfortunately, when our healthcare system is predatory, suspicions of motives easily spill over into biomedical science as well.

I don't know how to fix healthcare. I do know that alternative medicine is a good topic for moving classroom discussions away from just matters of truth and falsity, introducing questions about the social consequences of science and technology. And then, the debates get a lot more complicated.

It's Social, but Is It Science?

If natural science is our model for reliable knowledge, social science can appear less successful in comparison. Critics charge that "social scientific inquiry fails to test its hypotheses against empirical evidence even where it is possible, infects its theories and testing procedures with ideological assumptions, and conducts social inquiry within a framework of unrealistic and untestable hypotheses about human nature."[32]

Some influence of ideology is understandable. Most social science has an applied orientation; researchers want to solve social problems. Any theory addressing such problems will naturally come close to ideology. But from a natural science perspective, social science can look odd. On one hand, there is too much theory: multiple clashing, ideologically loaded schools of thought. On the other hand, while prevailing accounts of social mechanisms lead to plausible narratives, they are hard to test. As a result, the social sciences have too many resources to help construct reasonable stories after the fact. The sort of rationalizations and excuses for failure that characterize defenses of weirdness can flourish.

One possibility is that this state of affairs is unavoidable: the sort of precise mathematical relationships or clear causal structures typical in the natural sciences just don't exist in the social realm. The social sciences study insanely complex combinations of insanely complex objects. With increasing complexity, as we move from physics to biology, we begin to see more fuzziness. Biology can at least anchor itself on chemistry and physics, but psychology, which would be the natural anchor for social science, is itself insanely complex and frustratingly fuzzy. Social science can provide important insights on the workings of gender, ethnicity, or class, but such categories are fluid. They are constantly reconstructed and reconfigured, and are not always reliable bases for collective behavior.[33]

None of this, however, should prevent solid empirical work that establishes some basic facts. Statistics can help social scientists find relationships in the data. These relationships will often be more superficial than deep causal structures, but they will be useful nonetheless.

I would hesitate, however, to say that social science needs a fancier mathematical apparatus. For example, social scientists often work with qualitative data: survey results on a scale of "strongly disagree" to "strongly agree," data sets that summarize expert opinions on the level of democracy in countries or the extent of moral involvement of the gods in various religions. It can be tempting, then, to code such information as numbers and turn statistical methods loose on the data. However, it is not entirely legitimate to average, or otherwise statistically manipulate, numbers that are not genuine quantities but only indicate rankings. Unless done very carefully, analysis of qualitative data can produce false precision and misleading results, contributing to the replication crisis in social science.[34] Math in social science can be as much a problem as a solution.

Among social scientists, economists might say that they escape such criticisms. After all, they work with genuine quantities, expressed in dollars and cents. And they have far more detailed mathematical models and better statistical firepower. Indeed, the sophistication of economists is such that many of their models and theorems are of mathematical interest beyond the field of economics.

Now, I must confess that I have been suspicious of economics since I was an undergraduate, when I was required to take two semesters of economics. It was hate at first sight. The textbook was heavyhanded and unconvincing, and the math was window dressing. I aced the exams, but I stopped going to lectures and I sold the textbook as soon as I was done. I remain skeptical. After decades of market fundamentalism and being pummeled with praise of meritocracy, frictionless markets, and the virtues of privatizing and putting a price on everything, I have reacted badly. Today, I want to shrink the free market down to a size where it can be drowned in a bathtub.

So my students might call me biased about economics, and they could be right. Nonetheless, there are plenty of criticisms of mainstream economics, and some are relevant to weirdness.

Many critiques of economics target the goals of economics as an applied science. Economists often aim to maximize measures such as GDP growth by making economic processes as efficient as possible. The impulses behind such a goal vary, but they often include equating growth and efficiency with human progress and a kind of entrepreneurial moral individualism. That is not the only possible political perspective. Some critics argue that GDP is a terrible indicator of progress, although it is easier to measure than a mushy alternative such as happiness. Mainstream economics does not take environmental degradation into account, and fails to value care work and social reproduction. Economists gloss over imbalances in bargaining power in the labor market, treating the need for employment as if it were a consumer choice. Economics fails to distinguish between productive activity and parasitic rent-seeking in the financial sector. It discounts all social strategies other than that of an omniscient psychopath.[35]

Even if all these criticisms were correct, however, it would still be possible to apply economic knowledge to different political goals. Environmental economists propose to recognize costs ignored by current market mechanisms; feminist economists try to properly value care work. Some economists may carry their heresy so far as to favor labor unions to balance bargaining power. They might explore how to structure an economy to be more resilient in the face of events such as pandemics, instead of prioritizing a fragile efficiency.[36]

All this, however, is a matter of politics and goals rather than knowledge. Mainstream economics is suicidal in the long term, encouraging fantasies of perpetual exponential growth. But it still may provide an accurate description of how modern economies work, and identify the right policies for short-term GDP growth.

There are, however, also some criticisms that challenge economic claims to knowledge. Some philosophers of science observe that some of the core ideas of economics, such as neoclassical economic theory, have many suspicious features that characterize fake science. And there are heterodox economists who describe the economics mainstream as being stuck in a prescientific state that is often disconnected from reality.[37]

I think there is some merit to such charges. Much of economics remains centered on general equilibrium models with a single agent—the usual omniscient psychopath—representing all economic actors. While the math is attractive, people are not identical omniscient psychopaths, and many assumptions in the models are neither realistic nor do they have clear empirical validation. Model-building in economics is not always like the controlled process of simplification and approximation checked by experiments that is the engine of knowledge growth in physics. Moreover, the insistence on equilibrium is peculiar: from a physicist's perspective, economies do not look like systems hovering around a state similar to thermodynamic equilibrium. In equilibrium models, some of the most consequential economic phenomena, such as recurrent financial crises, only show up as external shocks. Finance, if it appears at all in such models, is treated such that finance-driven instabilities do not appear on the horizon.[38]

Economists, supremely confident in doctrines such as efficient markets, underestimated the likelihood of financial crashes such as that in

2008. They contributed to the fragility of economies in the coronavirus crisis of 2020. Such events provide an opening for critics, but nothing seems to change. The market fundamentalism of mainstream economics appears to be remarkably insensitive to empirical failure.[39]

Heterodox economists respond with their proposals for reform. But partly because of the ideological entanglements of the debates, and, more fundamentally, because professional controversies in economics are beyond my expertise, I find these proposals hard to judge. It's much easier to point out the shortcomings of mainstream economics than to construct a compelling alternative.

More encouragingly, there are also some signs of change within the economic mainstream. Some economists acknowledge that their discipline has become too insular, and that it needs closer connections to the other social sciences. Behavioral economics has forged closer ties to psychology, though it still tends to take deviations from the omniscient psychopath as failures of rationality. Some economics students have rebelled and demanded more pluralism in their education. It would be good if students were to learn from better textbooks than the one that started my unfortunate relationship with economics. And some economists worried about the reputation of their discipline argue that economics is changing, and that there is a difference between the cutting edge and the established views that remain so politically dominant.[40]

At present, however, the establishment controls the most prestigious journals and the cushiest academic and professional appointments. It is closely entangled with political power. It even has a fake Nobel Prize. This brings up an interesting contrast with the sort of weirdness that challenges science from outside of academic institutions. Fake sciences, I think, usually have institutional features that promote apologetics rather than learning. In cases such as intelligent design or alternative medicine, this translates into exclusion from the mainstream. The institutions of biology treat creationism with contempt; medical institutions intervene politically to deny alternative medicine market share. In such cases, our social mechanisms allocating prestige and power favor mainstream science. But if economics is a fake science—I don't think it's a clear-cut case—its position is inverted. Being close to power, being useful to the powerful, produces different forms of institutionalized distortions of learning.[41] And mainstream dominance alone does not automatically mean trustworthy knowledge claims. After all, practices like Freudian psychoanalysis, a reasonably clear-cut case of a fake science, used to enjoy mainstream credibility.[42] When my students have to evaluate claims to expertise, I don't want them to uncritically accept every institution that enjoys power or prestige.

If too many of the institutions of economics are structured to promote ideology, that problem seems common to many social sciences. And yet, partly because of their lower funding and influence, I find that I appreciate the social sciences outside of economics more. Due to insane complexities, I will never be able to trust a sociological explanation as much as the theories I am used to from physics. But sociologists work in ordinary academic settings, producing a form of expertise I usually trust. I have learned a lot about paranormal and religious beliefs from sociologists and social psychologists. I can't extend the same trust to the varieties of economics close to political power.

Facts and Meanings

Science is supposed to be about facts, not ideology. The facts are one thing, what they mean in the context of our lives another. It's tempting to think of science as an extension of our everyday perception, taking in facts beyond our usual capabilities, the way we can't see X-rays with our eyes but can detect them with our instruments.

But perception does not just deliver us facts. Indiscriminately absorbing facts about the world would not make any biological sense. After all, detecting information and processing it require resources, and reproductive success depends on not wasting resources. Perception and attention have to be selective. At every stage, an animal is best off picking out what is meaningful in terms of survival and reproduction and discarding the rest. Even if some chance variation had given us X-ray vision, maintaining that capability would have been a waste, since there is very little X-ray radiation in our pretechnological environment. What we perceive is always filtered through meanings, and these start with biology before culture.[43]

With culture, we get new layers of shared meanings and purposes. And historically, scholars—culture experts—have not separated facts and meanings. The Western philosophical tradition used to be far more concerned about practical wisdom and questions about a good life; it had much in common with spiritual wisdom traditions and was less focused on the abstract problems typical of today's academic philosophy. And various literary traditions from around the world, whether centered on epic poetry or religious devotions, have always been devices for reflection on meanings and practices, which have been inseparable from doctrines about the facts.

So my English major who was suspicious about science has plenty of company. Science has come to require a narrowing of focus, an emptying of the world of meanings and values so that we can have just the facts. But much of academic scholarship, particularly in the humanities, is still about more than the facts. Indeed, students of the humanities may justly say that they represent the main historical current of intellectual life. Today, all academic disciplines are under pressure to justify their existence in economic terms; English majors continually face the question of what their degree is good for. Science does not attract as much scrutiny, since if nothing else, it is infrastructure for technology and business. But the humanities can still help us to reflect on what all our power and knowledge is good for, and suggest ways of living that point beyond mindlessly accumulating wealth. Perhaps science and technology provide technical services, while the humanities address the important questions about meanings.[44]

I can accept much of this. I will not dispute that thinking about a good novel is more relevant to what most of us care about than solving a problem in quantum mechanics. I can even agree that the humanities should be central to our scholarly enterprises, not just producing their own forms of knowledge, but making knowledge humanly meaningful and imagining contexts in which the knowledge produced by science can take on varying meanings.

Some of us, however, do care a lot about the facts. And to get the facts and deliver its technical services, science does require a narrowly focused form of inquiry that strips away concerns for meaning as potentially misleading forms of ideology. Science aspires to a view from nowhere, to produce facts that remain solid regardless of how people may give them meaning.[45] The humanities do not add unnecessary layers of meaning onto facts; meanings are biologically and culturally inescapable. Instead, science performs an artificial, counterintuitive subtraction. And subtracting meaning would not have made sense except for the success science has enjoyed in producing facts.

That success, however, should prompt a conception of the humanities that takes science more seriously. Since facts and meanings have been so interwoven in both perception and culture, it has been natural to think of meanings as being objectively out there, independent of what people think. They are not. Meanings depend on particular biologies, social positions, and cultural commitments. Without facts, there are no meanings, but meanings are as much invented as discovered. Meanings depend on how facts are received as well as the facts themselves. Hence the humanities are full of unsettled interpretations, rival schools, and difficulties in achieving consensus. The humanities may still deal with important questions, but they have difficulty producing answers acceptable beyond particular cultural contexts. Calling poetry a deeper form of truth may be meaningful, but it is not truthful.

Compared to natural science, then, the humanities appear as a blend of knowledge and invented meanings where meanings have center stage. If we are looking for just the facts, what should we make of the humanities?

Consider history. Though usually lumped in with the humanities, history is also similar to natural science in the way it is concerned with establishing the facts.

I've done some work on the history of weirdness, and my wife is a historian. Every now and then, we co-author an article. I find that historical explanations require careful attention to evidence and a continual weighing of how new ideas fit in with or challenge already accepted explanations. This is much like natural science. Digging up new documents from archives is analogous to observations, and interpretations of such evidence are theories. Significant events in the past cause multiple effects, which can be independently traced back to the original events. Therefore, historians, like scientists, look for the convergence of independent lines of evidence, and build up a network of mutually supporting items of knowledge. Making up plausible-sounding stories is not good enough.[46]

Once, when I was complaining about the math anxiety I see among students learning physics, a historian colleague told me that he regularly encountered history anxiety. Historical reasoning is not intuitive or automatic; the demands of properly obtaining and using evidence can intimidate students who are not well-prepared.

All this means that I have no difficulty thinking of history as a knowledge-producing enterprise that is continuous with natural science. But together with the facts, history is very often entangled with producing meanings. If, for example, historians focus their attention on the role of slavery at the beginnings of modern economies, this can be because of new evidence, new explanation proposals that raise new questions, or also because of social developments that make questions about slavery more salient.[47] Like the sciences, history can have close connections to applications. The most public application of history tends to be efforts to draw the lessons of history for present concerns. And that naturally draws history into debates about meaning and ideology.

I don't think the entanglement of history with meanings should disturb those of us concerned about the facts, unless it leads to institutionalized distortions of history. Most of the time the weirdness and fake history historians face melts into a background noise of conspiracy theories and obviously crazy notions that space aliens built the pyramids or that the Middle Ages never happened.[48] But when historians operate in highly nationalist or conservatively religious environments, even academic institutions can easily be influenced by the need to validate prevailing meanings.

Still, if history comes close to the sciences in sustaining a stripped-down focus on the facts, maybe it is misclassified among the humanities. After all, how we cluster our academic disciplines is something of a historical accident, the way the classical liberal arts grouped arithmetic, geometry, music, and astronomy together under the quadrivium. Perhaps history has more in common with the social sciences than the literary humanities.

However we classify disciplines, though, it is hard to criticize efforts to construct meaning as being unscientific. Not every intellectual effort has to be directed toward getting the facts. The forms of weirdness that attract my attention and that my students explore in Weird Science often pretend to be scientific, to correct misguided forms of knowledge endorsed by a benighted scientific establishment, or at least to unveil some new and exciting facts. But when Freudian psychoanalysis migrates from being a failed applied science to becoming a tool for literary interpretation, it is no longer the same sort of weirdness.[49] Fundamentally, I am a math and science type, and unless they get the facts blatantly wrong, I find it hard to criticize meaning-making enterprises from a scientific perspective.

I am not always happy with my colleagues in the humanities. Their more postmodern exertions seem ridiculous, and their legitimate interests in questions of meaning and justice too often degenerate into a puritan moralism.[50] When that happens, I am glad I am on my side of campus, and I just hope the humanists won't influence the university administration to produce new bureaucratic headaches to complicate my life. Whatever grumbling I might do, however, is not about the facts, but the meanings of what we do.

The Continuity of Science

In the mid-twentieth century, philosophers of science sought a generally applicable logic of science. They tried to map out a scientific method, excluded weirdness by criteria such as falsifiability, and identified unscientific claims to knowledge within our intellectual traditions. All this implied a unity of science, even a unified vision of all knowledge. Scientific method produced reliable knowledge across all domains. In principle, all this knowledge could be translated into the language of the more fundamental domains. Though not feasible in practice, history could be reduced all the way down to physics.[51]

Philosophers have mostly given up on such a unity of science. The social and historical sciences were always difficult to assimilate into the natural sciences, and the quest for a common logic of science got bogged down and then abandoned. Today, instead of unity, pluralism reigns. We have lots of academic tribes that stake out specialized areas of study, all with their own methods and standards for acceptable knowledge. There are overlaps: physics and chemistry gradually shade into each other, going through chemical physics and physical chemistry. But by the time they reach history, physicists will be in a foreign country, where the practices of knowledge production are difficult to comprehend without long-term residence.

One reason for pluralism is that while our knowledge in different domains should not conflict, such constraints are very weak. Consider an economic transaction. Knowledge of physics is almost useless for understanding the exchange. Physicalism may be correct, so that any existing economy is made up entirely of physical objects and interactions, described by a combination of chance and necessity. Nonetheless, physics merely enables systems such as an economy to exist, without saying much of interest about how they work. The social structures and functional relationships that define an economic exchange can be realized with a vast range of physical substrates, even with physics very different than the fundamental physics of our universe.[52]

All that the existence of economies tells us about physics is that our universe must be able to support highly complex structures. We already knew this. The constraints physics place on economics are similarly weak. I sometimes have my physics students take a constant 3% economic growth rate and assume that this means a 3% annual growth in the volume devoted to sustaining human lives and activities. Then I have them calculate how long it will take for the radius of human-occupied space to grow faster than the speed of light. The answer turns out to be a few thousand years, which surprises them.[53] But then, we already knew that permanent economic growth was a crazy idea.

When the constraints are so weak, the older notion of a unity of science and exact translations between the sciences becomes unworkable. Our different domains of knowledge proceed by identifying their own objects of interest and their particular relationships.

Weak constraints, however, do not mean that anything goes. Humans are not exempt from biological evolution, and economies cannot violate physical limits. The domains of our various sciences overlap, and when they do, there is always work to be done to link up disciplines in ways that are acceptable to experts in all the disciplines involved. We can still have one network of knowledge which extends across all domains. An old-fashioned unity of science may not be feasible, but we can still recognize the continuities between our various forms of knowledge.

Examining weirdness often demands interdisciplinary work that exploits just such continuities. Weird Science students from all backgrounds find that they have something to contribute.

My collaborations with my wife include a case where we debunked a weird claim about medieval Muslim inventions of flight. In many Muslim countries, and in some fake-historical material intended for the consumption of Western Muslims, it has become popular to claim that Muslims had already invented some of today's major technologies. There are accounts of a ninth-century Andalusian, Ibn Firnas, achieving flight. A similar account from the Ottoman Empire in the seventeenth century has Hezarfen Ahmet Çelebi flying for about two miles across the Bosporus. In some Muslim countries, these claims are widely accepted as true. But the flights almost certainly never happened.[54]

Such claims of flight with artificial wings are physically impossible, especially when they imply muscle-powered flight. Comparison with modern gliding technologies shows that the claims are extremely implausible even as unpowered flight.

The historical documentation for the flight claims is very poor. There are only one or two very brief mentions of both the Andalusian and the Ottoman flights, in sources that date from centuries after the alleged feats, or that are travelers' tales also full of other sensational but unreliable accounts. Whatever may have been behind the stories of flight, they left no other trace, and no continuing practices. They certainly have no connection with the later development of flight technology.

Together with criticism based on physics and history, an awareness of common themes in weirdness also helps. Both Ibn Firnas and Hezarfen are said to have used materials such as eagle feathers. This suggests magical practices rather than technological tinkering. And beliefs about the Muslim invention of flight are institutionalized and accepted even by many academics in Muslim countries. It is not difficult to see how a religious and nationalist desire for prestige has shaped endorsements of fake-historical narratives.

Identifying weirdness does not require a common scientific method or a robust unity of science. We depend, instead, on criticisms rooted in the practices of particular disciplines such as history and physics. In the case of medieval flight, these criticisms converge and mutually support one another. Together with insights from other encounters with weirdness, we end up with a strong case for the stories of flight being fictions rather than descriptions of real events. Building such a case relies on overlaps and continuities in our forms of knowledge; examining many such examples strengthens the overlaps. And if history and physics had not converged, we would have been left with interesting questions for both disciplines. If physics had not cast doubt on human-powered flight or gliding with eagle feathers, historians would have had to assess the flight stories as somewhat more reliable. If there had been reliable documentation of the flights, physicists would have to take the occult powers of eagle feathers slightly more seriously. We would have to do some reconstruction on our network of knowledge.

Criticism of weirdness starts with particular mistakes in the context of disciplines, but it often spills over. There are common themes to many kinds of weirdness. For example, many forms of weirdness don't account for how knowledge of the weird is supposed to be obtained, resorting to occult means of knowing. This is what makes me suspect a Platonic metaphysics of mathematics—it portrays math as akin to mystical illumination or specially granted knowledge of conspiracies. There is only a limited repertoire for those who want to protect shaky claims from criticism, or who want to create a false appearance of science.[55]

Studies of weirdness, then, often have to cross the boundaries between disciplines. And both for my students and for myself, this is a large part of what makes weirdness fun. If nothing else, it's a great excuse to read books about just about any subject and pretend that it's research.

The Pursuit of Happiness

There once was a young woman who inherited a large fortune. She was in college studying psychology, which was fine, but not entirely satisfying. She then realized that her purpose in life was as simple as it was ambitious: to discover the secret to happiness. And now, with her inheritance, she had the means to find out. She dropped out of school.

She heard of a great teacher who lived in a cave on a remote mountaintop. After a lot of effort and money, she figured out the exact mountain and made the difficult journey to reach the cave. From the teacher, she learned that she needed self-knowledge, and that this knowledge was but the first step in changing herself for the better. Life was defined by suffering. To conquer suffering, she needed to overcome her cravings and her attachments to material things. She would then come to regard all from a perspective of detached compassion. Eventually she would embrace the Emptiness of ultimate reality, reach Enlightenment, and be free of all suffering.

All of what he said, however, were mere words, with shadows of meaning that merely pointed to the Path that she had to follow. She needed to put the words into practice and to walk the Path. Fortunately, the techniques needed to start her practice were described in a book the teacher had authored, together with its accompanying study guides. He would be happy to autograph the book, and he also highly recommended a series of affordably priced training videos.

So the heiress started out on the Path, dabbling with the techniques and practices. She made some progress, but then realized that her spiritual journey was starting to feel like university all over again. She wasn't convinced that being cleansed of attachments to worldly things needed to be such an involved process. Maybe in technologically less-advanced times this was the best that seekers could do, but today, shouldn't there be some precisely targeted drug or exact brain surgery that would do the job? And even if such an intervention was possible, was it what she really wanted? The end goal of the Path was supposed to be extinction and freedom from rebirth, but that sounded like annihilation. Death would take care of that eventually. The whole business of freedom from cravings was uncomfortably close to mere numbness to suffering. She didn't want a zombie-like state, even if it was called universal compassion. No, she wanted to be happy as *herself*, not to be transformed into some alien Enlightened being.

She then threw away her books and study guides, and went back to her quest for happiness. She heard of a monk who might be able to help her. After much effort and expense, she located the monk in a bare cell in a crumbling monastery deep in the desert. After much silence and prayer and looking deep into her eyes to grasp her soul, the monk announced that, like many young Americans, the heiress was too self-centered. True joy and contentment came from obeying the divine, and the divine will is that we should love and serve others.

Reorienting herself to service would not be easy. It called for some serious changes. But the Church would help and support her every step of the way, provided that she regularly donated to the Fund to Maintain the Shininess of Golden Ritual Objects.

And it came to pass that the heiress tried to regularly wash the feet of the poor and go to services and all that. It didn't last. She found that the familiar boredom was starting to creep back in. Once again, the whole self-transformation thing was not convincing. She wanted happiness as herself, not to become a saint. She could see that many of the believers found joy in participating in the community of faith, even in the midst of their poverty and hardship. And yet, that was not what she wanted. Fundamentally, it was not the American Way. The pursuit of happiness, as the generations before who had accumulated her inheritance had agreed, required acquiring lots of money. But the money was only a means to an end, which was to find happiness your own way, as a sovereign individual. Now that the money part was taken care of, she needed to figure out happiness.

She therefore sought out an expensive psychoengineer, who promised results through the latest and best science and technology. At first, the heiress was skeptical. Back in college, she had learned something about the psychology of happiness, and happiness research had struck her as a silly, barely scientific fad that was only a notch above the self-help literature you could pick up at airports. The psychoengineer, however, was able to assure her that a lot had changed during the years that she was exploring the paths suggested by the teacher and the monk.

Then there were consultations and brain scans and enormous amounts of data obtained through measurements with lots of fancy equipment. After all that work, the psychoengineer determined that while the heiress was highly intelligent and capable, she was also fundamentally a shallow person. Even back in college, she had most profoundly enjoyed her partying rather than her studies. She had just wanted to think of herself as having hidden depths of character. But in her case, far from being the key to happiness, self-knowledge was an impediment. In fact, even a lot of her knowledge about the world was an obstacle to her goals. Her bliss would require an appropriate measure of ignorance.

Fortunately, given the heiress's wealth, the problem was eminently solvable. The psychoengineer proclaimed that for an enormous fee, she could remove all the barriers that kept her client from being happy. The heiress needed a life of sex, drugs, and rock and roll, with particular emphasis on custom-designed chemical aids and a constant stream of diverting entertainments. The full program for happiness included a large staff to isolate the heiress and to plan her affairs, preventing her from distressing encounters with anyone far outside of her own social class. She would be exposed to nothing but luxury. To keep her secure in her long-term enjoyment of her private paradise, just the usual army of wealth-management professionals and flunkies would not be sufficient. The heiress would also be equipped with state of the art artificial intelligences dedicated to insulating her from any intrusions that could dampen her bliss. She would become a pampered brain cocooned and dedicated to happiness.

The heiress paid the psychoengineer and gave orders for the project to begin. In less than a year, the construction of a Pleasure Compound and Fortress was complete. The heiress walked in, never to emerge again.

And she lived happily ever after.

5 REASONS TO BELIEVE

Weird Experiences

A common reason for belief in the paranormal is a weird personal experience.[1] So I expected my Weird Science course to yield a wealth of compelling stories of magical occurrences and supernatural encounters.

This hasn't happened. I get accounts of weird experiences, but they are rarely spectacular. Even students who believe in the paranormal can be reluctant to speak about their experiences in class. They loosen up in the papers they write, but even there, most of their stories are full of ambiguities and uncertainties. If they think that ghosts are real, and they have a personal anecdote that has helped make up their mind, they still are usually tentative about the strange sights and sounds they relate.

Therefore, I often have to tell my own stories to get the classroom discussion going. The most impressive tale I have comes from my days in graduate school.

On a Sunday morning, I woke up and noticed my wife standing next to our bed. She was not, however, moving: she was gazing out into the distance without noticing me at all. I started to get up to ask her what was happening. After all, she was supposed to be away at a history conference. She was to be flying back later. But as I attempted to get out of bed, I found that I could not. I was completely paralyzed, and I had an agonizing few seconds as I attempted to fight my condition. I snapped out of it and got up to find that I was alone in our apartment. I then experienced a bizarre state of mind. Sentiments of doom rushed over me; I wondered if I had had a premonition of her plane crashing. It took me a few more minutes to start to settle down.

Curiously, I had read about the phenomenon of sleep paralysis a few months before the incident. Our bodies are paralyzed during dreams, so we don't actually perform the muscle actions we dream about. And clearly my sleep paralysis mechanism had been occasionally malfunctioning: every now and then I woke up because I kicked a ball in my dream and actually swung a leg. The base of the lamp on my side of the bed was broken due to my swinging an arm at it during sleep. According to my reading, sometimes sleep paralysis overlapped with half-waking consciousness for a short time during the process of waking up or falling asleep, and these periods could also have accompanying hallucinations.[2]

I find it interesting that even with my ordinarily skeptical inclinations, my first, almost automatic thoughts were to interpret the experience in paranormal terms. What happened must have been a hallucination, but it felt as real as any everyday event. And for the first few minutes, I really was anxious that a disaster might have taken place and that my vision of my wife was somehow our last contact.

I still occasionally swing a limb in bed, but I only had one more full-blown episode of sleep paralysis and hallucination, soon after I started teaching at my current university. This was when I was about to fall asleep, and it was a classic night-hag experience.[3] I was paralyzed and felt attacked by a malevolent external presence on top of me, pushing down on my chest. And once again, although I knew even more about what the psychological literature said, it took me a few minutes to shake off my peculiar state of mind. During that period, I was half-convinced that I had encountered a paranormal being.

I can see how weird experiences can generate paranormal beliefs.

The experiences need not even be firsthand. A friend once told me that he thought that ghosts were real because his grandmother, his favorite relative, had told him about her encounters with ghosts. He acknowledged that a belief in ghosts was odd for a physicist like him, and that he had no evidence that would carry any weight for an impartial observer. But he was simply unable to distrust his grandmother.

Now, sleep paralysis is reasonably well understood. As with anything involving insanely complex brains, there is still much to learn about hallucinations. Still, brains can adopt flawed models of reality and make odd inferences in circumstances with ambiguous sensory input under unusual conditions.[4] Doubters of the paranormal, whether the skeptics among my Weird Science students or readers of *The Skeptical Inquirer*, are usually confident that psychological knowledge will deflate weird claims. In fact, most skeptics think that paranormal belief is irrational. If I were to believe in premonitions and night hags immediately after my weird experiences, that might be understandable. But that would be because I was not sufficiently informed. If I were to do my homework properly, I would run into the psychological literature and find out that the better explanation was sleep paralysis and hallucination episodes. If I were to then dig my heels in and insist that I encountered a night hag, a skeptic would further dissect my errors in reasoning, identifying my informal fallacies for my benefit and general annoyance.

I usually have one or two psychology majors in Weird Science, and they tend to be skeptical. I never have to explain confirmation bias or the reconstructed nature of memories to the rest of the class; I just ask one of the psychology students to do so. Clinical psychology has its own history of weirdness, from the earlier schools of Freud, Jung, or Reich, to more recent screw-ups involving recovered memories or multiple personalities.[5] Still, even my therapy-oriented psychology majors seem to have learned not to uncritically trust weird experiences.

If my students get too skeptical, however, there won't be any interesting discussion. So sometimes I have to criticize their casual skepticism. For example, I discourage students from diagnosing fallacies associated with paranormal claims. Most so-called fallacies actually work well in everyday situations, and any notion of rationality that attempts to reduce reasoning to a list of rules will self-immolate in short order.[6] More important, it will not do to shoehorn paranormal belief into a set of cognitive and reasoning errors categorized by psychology and informal logic while associating skepticism with an abstract ideal of rationality. Skeptical beliefs are no less rooted in experience and in brain processes with limits and blind spots. Rationality is not a magical power; it has a psychology. It has costs as well as benefits.

Examining weirdness should also prompt us to look critically at processes that constitute rationality. Skeptics think of paranormal belief as a failure of reasoning, but in some circumstances, psychological processes that we associate with deliberative, rational thinking may well support false beliefs. Ideals of rationality impose costs upon us, and sometimes these costs may be unreasonably high.

Sources of Weirdness

The anomalistic psychology textbook I occasionally consult lists many correlates of paranormal belief. There is no single dominant theme. Socioeconomic status and education, even science education, does not seem to matter much. A lack of security in life circumstances weakly correlates with religiosity, and a similar pattern also holds with less organized paranormal beliefs. Like a taste for music, paranormal convictions appear to be universally present in human societies, but they take on varied forms, and the distribution of these forms does little to explain the sources of weird beliefs.[7]

Intuitive ways of thinking, which are fast and have low cost, but are more error-prone, tend to support paranormal belief. People inclined to perceive illusory patterns in haphazard data are also more likely to believe in weirdness; in fact, a number of well-known cognitive biases seem to help produce religious and paranormal belief.[8] During a night-hag experience, a person may intuitively feel that they have encountered a supernatural evil. But later, reflecting on and reasoning about the experience, they may come to think otherwise. Such evidence fits a model where paranormal beliefs arise from glitches in perception combined with deficits in critical reasoning. Still, none of these psychological tendencies are overwhelmingly strong. There must be more going on.

One question is why weirdness only takes particular shapes. The weird belief that a small number within the human population are actually aliens or alien-human hybrids is not wildly popular, but it exists and propagates.[9] More important, even if they had not yet encountered this form of weirdness, when they do, my students recognize that some people might in fact believe that aliens are among us. But if I were to say that a similarly small number of people have their lungs turn into glass-enclosed cavities every other Tuesday that host cage fights between miniature purple orangutans, my students would immediately know that no one actually believes this. Some claims are weird but will still find adherents or be entertained in fiction. Some claims are just crazy.

Human minds, therefore, must be structured to favor some weird beliefs and not others. Moreover, since human brains are a product of evolution, cognitive scientists have some tools to help figure out how such a structure came to be.

The past few decades have produced some very interesting explanations of our propensities to accept paranormal ideas, especially supernatural agents. These explanations often start with the observation that the most important and complex features in our environments are other humans. If we fail to detect the presence and intent of a human agent, such an error is likely to be more costly than mistakenly detecting agents when none are present. Therefore, evolution has favored mechanisms in our brains that automatically recognize human faces, even though these mechanisms overperform and we regularly see faces in the clouds or on pieces of toast. It is plausible that our brains have additional mechanisms that make us very readily detect the presence of agents and purposes, even when they are not around. In conditions of uncertainty, we naturally jump to the conclusion that there are human or human-like agents responsible, whether these are ghosts or conspiracies.[10]

Normally functioning human minds also automatically classify the objects we encounter into categories that are most meaningful for apes like us: people, animals, plants, tools, and inanimate objects. Each of these categories comes with a rich set of inferences associated with it. If we detect a person, for example, we immediately infer that she has purposes, was born, can talk, and so forth. A statue, in contrast, has none of these properties. Interestingly, the stories we tell, whether in fictions or popular religious tales, are full of objects that are weird but not too weird. The notion of a talking statue or a human who is really an alien strains the basic categories and thereby draws attention. But the violations of the inferences associated with the categories are minimal in these cases, so a talking statue or an alien in human guise are concepts that can be easily remembered, reasoned with, and propagated in stories. People with occasional glass lungs housing miniature orangutan fights are too hard to remember, and such concepts too badly scramble up the inferences that come with the human category. They will not appear in either fiction or religious stories.[11]

Combining a tendency to perceive agents with the memorability of weird agents brings us close to supernatural agents, but there is still considerable debate about the exact mechanisms that predispose humans toward paranormal belief. We might have some dedicated, modular structures in our brains, as some evolutionary psychologists propose. Alternatively, more general learning processes that have the brain acting as a predictive device might suffice in the right context.[12] In any case, it is plausible that the structure of human minds draws us toward some kinds of weirdness and not others.

There is more to the paranormal, however, than individual psychological tendencies. Weird beliefs are shared: they are socially affirmed and propagated. Stories need an audience. And human communication is full of reasons being offered and scrutinized. We are most readily impressed by reasons that can be checked, or are already accepted, or are validated by commonly accepted authorities. Even if communication is more about persuasion than achieving truth, getting others to accept bla-tant falsehoods is not easy.[13] It is not enough for ideas such as ghosts or aliens in disguise to be memorable and rooted in strange experiences; they must also survive the filtering inherent in everyday reasoning pro-cesses. Weirdness must become part of a shared culture.

Cultural evolution is another area where theories are speculative and difficult to anchor upon properly established brain processes. For example, there have been repeated attempts to draw a close analogy with biological evolution by postulating a meme, a unit of culturally transmit-ted information that plays a role similar to that of a gene in biology. I have a small stack of books on memetics, each of which define the meme very differently.[14]

Memes are a good reminder that cultural evolution need not ben-efit the hosts of ideas, and memes also focus our attention on the re-production of information. But culture does not propagate analogously to duplicating strands of DNA. When we communicate, we don't copy the exact wording and intonation of what we hear. We discard most of the detailed information and obtain the gist. Fidelity of transmission de-pends on similar structures in minds, so that communication can evoke shared representations, categories, and attendant sets of inferences. Cul-ture propagates and evolves in a manner that depends on this underlying structure.[15]

Whatever an eventual fuller explanation of paranormal beliefs will be, it must connect the evolved structure of minds with the social pro-cesses that shape culture. World religions, for example, have much more staying power than stories about aliens among us. It is plausible, though still speculative, to think that the success of world religions has a lot to do with deploying supernatural beliefs, such as morally concerned gods, in ways that promote large-scale cooperation beyond kin groups.[16]

Cultural evolutionary explanations, when fully developed, will apply to our accurate beliefs and scientific representations as much as weird-ness. *The Skeptical Inquirer* calls itself "the magazine for science and rea-son." Slow, deliberative reasoning processes also shape our landscape of beliefs. Intuition, used critically, will always have a place in science. Critical reason is more reflective, but it also depends on evoked inferences and the shared structures of minds.

Understanding reason as an evolved capability of physical brains and communities is important, because then we can recognize varieties of rationality with different costs and benefits. We can distinguish between the idealized descriptive accuracy that is the goal of science and reason as a practical tool. And then, we find that some kinds of weirdness draw on slow, deliberative processes as much as intuitive leaps. And some false beliefs will be rational in some circumstances.

Excusable Mistakes

Recently, the crumbling brick walkway leading to my front door had to be fixed. I had the options of working on it myself or hiring someone. Since I have no skill with bricks, I could only have produced a botched job that would have needed even more expense to set it right. The rational decision was to pay someone who knew what they were doing.

Instrumental rationality is about choosing the best means to our ends: achieving the best outcome at the least cost. There is more to instrumental rationality than the omniscient psychopathy of rational actors in economic models. Our interests are not always self-centered, and knowledge about our options and likely outcomes itself comes at a cost.[17]

Sometimes, as with my brick pathway, the relevant knowledge is cheap. I'm well aware of my incompetence with bricks. But knowledge is often costly. For example, in political science, there is a debate about the rationality of voting. In a democracy, we like to imagine well-informed voters weighing their options, comparing likely futures in light of their particular interests. Unfortunately, being well-informed about policy options is more than a full-time job. Even the details of our town's local water management are beyond me—or would bore me so much that I'd gnaw my own arm off to be able to return to thinking about weirdness. Since knowledge is costly, decisions have to be endorsed by uninformed voters. Moreover, the influence of each individual voter is tiny. Especially for low-income voters whose interests are ignored by politicians, the benefit of a vote for a lesser evil may not be enough even to offset the costs of the act of voting, much less the costs of becoming better informed.[18]

There are some workarounds. Political parties can help share the costs of knowledge between people with similar interests. If I vote in someone with whom I share an ideology, I could have some confidence that they will further our common purposes. Furthermore, organizing together with like-minded people will mean that we will share the costs of generating and evaluating policy options. I won't have to gnaw my arm off, only figure out which party will work best for my purposes.

The reality is a bit more complicated. Around the parts of campus where I spend my time, the dominant ideology is meritocratic liberalism. In other words, I am surrounded by status quo conservatives who care about diversity and inclusion, by which they mean that everyone must come as close as they can to becoming an omniscient psychopath, which will make sure that they can rise to their level of incompetence in a meritocratic hierarchy.[19] In the small town I live in, however, the meritocratic liberals are outnumbered by nationalist conservatives who resent being condescended to by wealthy moralistic liberal scolds, and who prefer to be led by even wealthier businessmen and moralizing preachers.

The winners in our local politics, then, tend to be religious and nationalist conservatives: right-wing populists. The liberal faction of conservatives still vote, out of a fossilized sense of civic duty, even when they have no hope of winning. But is political activity a rational use of their resources? I am not sure. In the short term, it is doubtful. They rarely win. But in the long run, they may think it best if liberal organizations don't completely atrophy, in case things change and they suddenly have a chance. The situation is murky: are efforts on behalf of a losing cause a waste of resources, or are they expressions of dedication and purpose, gambling on future successes? After all, for many, political identity inspires as much devotion as ethnic or religious loyalties.

The instrumental rationality of political actors can be hard to judge.

Still, skeptics usually think that resisting vaccination or believing in aliens among us are not murky cases. We still need to account for the costs of knowledge and the tangle of interests knowledge is supposed to serve. Otherwise, calling our favored actions "rational" would be little more than cheerleading. But we can do the accounting, and believing in weirdness turns out to be like my having a delusion that I have bricklaying skills.

Imagine some parents who want their child to be as healthy as possible. They consult some relatives and watch some YouTube videos, and then decide to refuse vaccinations. Later, a friend who is a medical doctor explains why vaccines are a good idea, so the parents waver. But before they change their minds, they watch another video recommended on their social media feed, and decide to rely on their gut feelings against vaccination.

This decision would not be rational. The parents' purpose, which is the health of their child, would be best served by accurate knowledge about vaccinations. This knowledge was available at low cost. The parents can be expected to do their due diligence and obtain the relevant knowledge, and then correctly reason about the right course of action. They have not done so. This is not a murky example. My students in Weird Science agree that refusing vaccines is irrational. They almost universally think that anti-vaccine beliefs are both thoroughly false and completely harmful.[20]

With vaccines, however, the costs of knowledge are low. We need to also look at cases where the costs of acquiring knowledge are much higher. Scientific medicine is fallible. For example, stomach ulcers were once thought to be due to stress or spicy food, but it turned out that they were often caused by bacterial infections. Medical professionals doubted the possibility of such infections, because they did not think that bacteria could survive stomach acids. Moreover, treatments such as reducing stress had independent benefits that fooled many doctors into thinking that stress caused ulcers.[21]

Imagine, then, that a patient seeks relief for a serious condition, but that the medical establishment harbors a false consensus on its cause. A few dissident doctors point out the errors in the received opinion, but the vast majority of medical professionals are caught up in a lazy consensus sustained by groupthink rather than proper reasoning.

In such a case, it would still be rational for a patient to take the consensus medical advice, even though she will then have false beliefs about her condition. Due diligence, for the patient, requires only that she has informed trust in the overall competence of the medical profession, and perhaps that she makes sure her treatment is confirmed by a second opinion. The knowledge that most doctors are mistaken about one particular ailment is far too costly to obtain: it would require her to learn enough medicine to approach expert status. Otherwise she would not be able to evaluate the detailed technical arguments of the dissident physicians. Without such knowledge, it is by far the safer bet to follow the expert consensus, although she has the bad luck to be wrong in this instance.

For such a patient, a false belief about her condition would be excusable.[22] Rationality does not guarantee accuracy. The high cost of acquiring knowledge can make a tradeoff that leads to lower costs and higher error rates reasonable. Science is a collective enterprise of lowering the costs of knowledge. For nonexperts, it is almost always a better bet to accept the expert consensus, unless they have good reasons to believe that the experts cannot be trusted. Dark matter may turn out not to exist, but for now, it is the best bet both for astrophysicists and anyone curious among the general public.

All this may seem like stating the obvious. After all, isn't it always best to seek accurate knowledge in order to make the best decision? There is a slight complication. The patient has an interest in accurate knowledge of her condition. But that is not the same as her interest in being healthy. Her interest in accuracy is strictly instrumental: accuracy is useful only insofar as it would help with better decisions regarding health. Otherwise, the medical details may be as exciting as local water-use regulations. Still, her interests align with each other. If the dissident physicians were to overcome the false consensus and reduce the cost of acquiring accurate knowledge, the patient would happily accept improved medical knowledge and adjust her treatment accordingly.

Therefore, it might seem that accounting for the costs of knowledge just adds a small qualification to skeptical accounts of weirdness. Weird beliefs are still due to breakdowns in reasoning. If weird beliefs persist, that persistence is still caused by the kinds of fallacies and biases that skeptics like to diagnose. All our interests will be better served with more accurate knowledge, and the advance of science represents progress for everyone.

Once again, it's more complicated than that.

Costs of Possession

Knowledge can be costly to acquire, but the possession of knowledge also imposes costs. In the health-related examples, the possession costs of accurate knowledge were negligible. Better knowledge could only improve health outcomes. If the possession of knowledge imposes heavy costs, however, accurate knowledge may no longer serve our interests.

Many denizens of my town responded to the COVID-19 pandemic by refusing to wear masks. Conspiracy theories were rampant, and though I couldn't gauge the local popularity of weird beliefs such as the virus being a hoax, many appeared to believe that the dangers were exaggerated and masks were unnecessary. To make matters worse, the prevalent right-wing populism already included a general suspicion of expert advice. For some, avoiding masks became a way of displaying loyalty to their political tribe. Local communities, especially those organized around churches, have always been networks of solidarity, providing the benefits of cooperation and lowering the costs associated with life. Signaling disloyalty, therefore, could have a high price: missing out on the benefits of mutual aid. Meanwhile, most university faculty, who were invested in trust in expertise and who were less attached to local community networks, displayed their liberal loyalties by wearing masks.

The costs of acquiring accurate knowledge were low: information about masks was easily available on the mainstream media. But the high possession costs of knowledge among right-wing populists influenced many to make poor decisions about masks.

Refusing masks was probably not rational. After all, the mask avoiders had a strong interest in their own health and the health of others in their immediate community. Accurate knowledge would have helped them further these interests. But the mask example also suggests that if false beliefs help define cohesive communities, there may be cases where the benefits of believing in falsehoods outweigh the costs. Religion and nationalism are ideologies that help organize large-scale cooperation and are notorious for endorsing false beliefs about science and history. Ideologically vital falsehoods are the best candidates for weird beliefs that are maintained by processes other than cognitive biases and reasoning errors.[23]

Imagine two countries, Turania and Urartia, where the main ethnic groups are Turans and Urarts. Over the centuries of their existence, both countries have been the centers of regional empires, and their rival monarchs have often made war on each other. The border provinces have changed hands many times. There have been periods where Turania has conquered Urartia, and times when Urartia has reduced Turania to a vassal state. Ethnic minorities speaking Urartic still live in Turania, and talented Urarts have risen high in the Turanian imperial bureaucracy in the past. Today, Turania has become a republic and Turanian nationalism is the popular and officially sanctioned ideology. The two main political parties represent different nationalist factions, regularly accusing each other of treachery or incompetence in furthering the Turanian national cause. Urart nationalism dominates on the other side of the current, disputed border.

Turanian nationalism has shaped what has become the Standard Story of Turanian history. The Standard Story is taught in schools and frames discussions of current affairs in the Turanian mass media. The Standard Story is not a complete fabrication: it gets the dates of major battles right, and does not take the legend of the divine parentage of the founder of the first Turanian dynasty too seriously. It entertains, but does not fully endorse, conspiracy theories about how a cabal of ethnically Urart officials betrayed the First Empire. The Standard Story is, however, a version of events where the Turanian emperors are noble and generous, while the Urart kings are perfidious. When the Turanian Empire ruled Urartia, it bestowed the greatest possible gift on the Urarts: knowledge of the True Religion. Urarts, however, were ungrateful, and came to favor a heretical version of the faith. Just over a hundred years ago, there were episodes of ethnic cleansing of Urarts in the Eastern Provinces. These episodes were provoked by Urart rebellion and disloyalty, and did not rise to the level of genocide—accusations of genocide are lies promoted by the Urartian regime to support its territorial claims. In fact, focusing on the ethnic cleansing of Urarts distracts from the criminal expulsion of ethnic Turans from what is now Western Urartia.[24]

The international community of historians consider the Standard Story to be a mixture of real history and Turanian nationalist myth, incorporating many falsehoods. Nonetheless, most academic historians working in Turania defend sophisticated versions of the Standard Story. Funding for history and decisions about hiring and promotion in academic posts are influenced by conservative donors and nationalist foundations. Not too long ago, a handful of Turanian historians who deviated from the Standard Story were labeled Urart sympathizers in the popular media and hounded from their jobs after a public uproar.[25] The professional environment of historians, therefore, selects for nationalist historians who sincerely believe in something like the Standard Story. The popular historical narratives available in mass-market books and shown on television programs take the Standard Story for granted.

Now, in this situation, imagine an ordinary middle-class Turanian citizen who, much like almost everyone she knows, has a strong but nonfanatical commitment to Turanian nationalism. Being of the Turan ethnicity and nationality is central to her personal identity. She affirms her loyalties among friends and helps reproduce nationalist beliefs in her children. She also receives material benefits from being a member of a cohesive Turanian nation, who support each other in times of need and sustain a common culture that reduces the costs of figuring out whom to trust. Turanians are able to take collective action to exploit weaknesses among the Urartians, and to defend themselves against Urartian hostilities when necessary.

As part of the nationalist package, our patriotic citizen accepts the Standard Story, including its falsehoods. Her cost for acquiring more accurate knowledge is high. She might step into a dissident bookstore or stumble onto a YouTube channel that provides alternative views, but these views have a bad reputation and are hard to distinguish from the cultish conspiracy theories that are also dismissed by mainstream pundits. She does not speak a foreign language, and she has no specialized training in history. Therefore, she is not able to properly evaluate dissident claims, and acquiring historical training would be too costly. Indeed, she has no deep interest in history. Her historical curiosity is minimal and only instrumental, only to the extent that it serves her primary interests, which are bound up with nationalism. She trusts the Turanian experts, and is well-satisfied with the mainstream media discourse which does not seriously question the Standard Story.

Moreover, there are steep possession costs associated with more accurate beliefs, since nationalists are sensitive to signs of dissent. The recent border incidents have intensified vigilance to maintain loyalty among Turanians. Like every community, Turanians have ways to detect free riders who want to exploit the benefits of cooperation without pulling their weight. On sporting events and national days, and often on lesser occasions, Turanians expect public signals of national loyalty. Some of these demonstrations, such as reciting patriotic poetry or singing anthems, directly refer to the Standard Story. If our patriotic citizen were to change her mind and join the dissenters, she would either have to conceal her views or pay the costs that come with ostracism: community shunning, a restriction of economic opportunities, and harassment of her children at school.

It might seem easy to conceal dissenting beliefs by singing along with the crowd and uttering formulaic praise for the national heroes of the Standard Story. It should be possible to suppress the urge to blurt out that the heroes were slavers and mass murderers, and instead to deplore any suggestion that their statues should be pulled down. Such deception, however, is costly. To start, there is the burden of constantly keeping track of both actual beliefs and the beliefs to be displayed in appropriate situations, and figuring out the appropriate emotions to be faked. As many a closeted gay person or cleric who has lost their faith but needs to go through the motions to maintain their livelihood can attest, living a double life can have a heavy cost.[26]

Affirming the Standard Story signals loyalty. But not just any story can do the same job. A serviceable story cannot be too easily doubted, like the divine parentage of the First Emperor. But as long as they remain plausible, even the falsehoods in the Standard Story can be useful. After all, only people committed to Turanian nationalism would be likely to absorb and affirm the Standard Story. Outsiders such as Urarts, or neutral parties who are just curious about history, would not be brought up with the Standard Story. The historical sources they learn from would not be Turanian, and therefore their sources would not endorse a Turanian nationalist pattern of falsehoods. So precisely because it is partly false, the Standard Story serves as a good signal of group loyalty.[27]

In the case of the patriotic citizen, then, both the possession and acquisition costs of accurate knowledge are high. Moreover, her interests are aligned with having the false beliefs embedded in the Standard Story. It is instrumentally rational, in her situation, to continue to believe in the Standard Story. It is not merely an excusable mistake, a good bet that turned out to fail. The best available beliefs that serve her interests include falsehoods.

Since the patriotic citizen's false beliefs are instrumentally rational, slow, deliberative thought processes will take part in maintaining her beliefs, as well as her intuitive deference to her community. If challenged, she may resort to motivated reasoning—she might become selectively informed by consulting sources considered reputable in a nationalist environment, and be satisfied by the way that nationalist historians purport to debunk criticisms of the Standard Story.[28]

While difficult to avoid, however, motivated reasoning still looks like a reasoning error. A more likely strategy, one that need not involve errors, is for the patriotic citizen not to deeply engage in any historical disputes. After all, she only has a low level of interest in history, and that is only due to the historical component of her nationalism. Her primary interests are to maintain her community and its cooperative advantages for herself, those close to her, and her children. Her identity and her long-term projects in life depend on the prospects for her nation; the Standard Story merely has a supportive role. Her interests are already well served by trust in the nationalist Turanian experts and their popular and schoolbook renditions of history. She has no reason to devote resources to any further involvement. She can outsource nationalist apologetics to a trusted set of experts, and not waste resources on inquiring too deeply about whether nationalist historical narratives are accurate.

Fully Reflective Rationality

It still seems odd that false belief can be instrumentally rational, even if the Standard Story serves our patriotic citizen's interests and is not immediately dislodged by reasoning. There may yet be an instability lurking in the example.

The patriotic citizen retains belief in the Standard Story by not being completely reflective about her belief-forming processes. Especially in academic settings, we usually demand a more reflective sort of rationality. Not only do we offer reasons for our beliefs, but we also expect to be able to stand back and reflect on our reasoning process. Is our reasoning good? Have we collected solid, representative evidence? Are our methods reliable? Do we have interests that conflict with obtaining accurate knowledge?

Such reflections help identify false beliefs. Conspiracy theories, for example, may be very well protected from direct refutation: any evidence against the Illuminati running the world must have been planted by the Illuminati, which just illustrates how powerful they are. But we can then ask how the conspiracy theorist happens to know about the Illuminati's schemes. Their response invariably comes down to secret knowledge somehow granted to a select few. In contrast to the sheeple, the conspiracy theorists are able to read the surface patterns of events and intuit the plot that is the deeper explanation. If so, the claim that the Illuminati run the world is no better than any arbitrary made-up claim. If it is correct, it is correct by accident, but that is very unlikely. The conspiracy theory is sustained by unreliable methods and distorted evidence.

In that case, if the patriotic citizen were to become more reflective about her nationalist beliefs, this could undermine her confidence in the Standard Story. For example, she could develop a more intense nationalism, and come to care deeply about demonstrating the merit of Turanian claims to any doubters. She would then enter a dissident bookstore and start researching what international historians have to say, intending to expose their errors. Along the way, she may well realize that much of the Standard Story is mythical. The risk she runs, however, is not just that her nationalism will become softer. Her nationalism may unravel altogether. In that case, her research will lead to a painful process of personal transformation, after which her friends will notice that something important has changed about her. She will scarcely be the same person anymore, holding back on national days, lacking enthusiasm when her children go on school trips to visit monuments to national heroes.

Reflection, then, introduces the possibility of not just a change in beliefs, but also an instability in the underlying interests the beliefs are supposed to serve. In cases such as nationalism, where beliefs and interests depend on one another, the framework of instrumental rationality cannot fully handle instabilities in interests. If it is no longer true that our interests are fixed, then we cannot just weigh courses of action according to how they might serve those interests.[29]

A more fully reflective form of rationality would avoid instabilities.

Reflective rationality does not guarantee accuracy, but errors will now be excusable, similar to taking doctors' advice that turns out to be mistaken. And since accuracy will now be an overriding interest, getting something wrong will not provoke an identity crisis but a routine adjustment in representations of reality.

Still, it is not that easy to recommend fully reflective rationality for everyone. Instrumental rationality naturally incorporates motivations to take action. If I have an interest in health, I should eat more broccoli and less ice cream. If I regularly give into temptation, I will be behaving irrationally, or, alternatively, my interests are not dominated by health but by short-term pleasure. If I never act to further my interests, observers would either question my sanity or doubt that I really have such interests.

Instrumental rationality, then, can accommodate a diversity of interests, has a natural connection to motivations for action, and readily incorporates costs, including the costs of knowledge. In contrast, fully reflective rationality seems most appropriate for fanatics who care for accurate knowledge above all other considerations. Why, in a world where interests vary widely and knowledge is costly, would anyone take on the extra costs of reflective rationality? What is the motivation?

Since interests can change as well as beliefs, an account of interests will be useful. As I see it, interests start with biology: with reproductive interests.[30] Successful reproduction, after all, is the explanation for the forms of life that exist. As highly social and communicative apes, we also have interests rooted in cultural reproduction. We sacrifice for ideas as well as our offspring. Interests such as health and acquiring material resources are tethered to the reproductive interests that health and wealth ultimately serve. Our interests can come into conflict: a desire to have and care for children and the duty to achieve martyrdom to further an ideal might not both be realized. We usually, however, settle into a reasonably stable configuration of interests.

Biological evolution works on variation and produces variety. Monocultures are not common, and the artificial monocultures imposed by modern agriculture tend to be unsustainable.[31] Instead, we find complex

ecologies harboring organisms with different evolutionarily stable strategies. We should expect much the same with cultural evolution: social
ecologies with multiple interdependent and successfully reproducing
ways of life. This means different stable clusters of interests and attendant
social identities. We have a diversity of ethnicities, sects, subcultures, social classes, and more. Individuals have a variety of ways by which they
locate themselves in their social ecology.

Such a varied ecology is full of conflicts of interest. We have opportunities for cooperation, but also occasions for exploiting one another.
Some of these conflicts can be dissolved through reasoning together. But
reasoned persuasion is only one item in our social toolbox.

Consider, again, the rival political coalitions in my town. Liberal
elites and right-wing populists conflict over managing social diversity.
Liberals favor a bureaucratic, explicitly rule-based business and professional monoculture that can accommodate incidental forms of diversity
of ethnicity and gender. Diversity, for liberals, is having lots of ethnic
restaurants with much the same forms of ownership, management, and
interaction with customers. In contrast, ethnic and religious nationalists
embrace competition between groups and seek to preserve their local
advantage rather than adopting the liberal monoculture. Where I live,
this manifests as nostalgia for the informal Christian establishment of
the past.[32] Meanwhile, liberals evangelize for a secular faith that has not
yet established full dominance over those it portrays as deplorables.

Since it more often aligns with my interests, I usually join in the secular liberal coalition. But I think the politics I find myself observing is
primarily a contest of power rather than a conflict between reason and
unreason. Furthermore, I expect that any political victory would eventually lead to an unstable situation where the winning coalition of interests
would fracture, leading to a new ecology of competing and cooperating
cultural formations.

In such a varied and changing social ecology, then, what are the
prospects for a commitment to accuracy and a fully reflective form of rationality? Everyone rhetorically favors accuracy. It is psychologically dif-

ficult for anyone to say that they don't care for truth, only what is advantageous. We engage in motivated reasoning, but we also try to maintain an illusion of reasoning properly.[33] It is psychologically almost impossible to say that a claim is false, but that one should believe anyway. Even leaps of faith don't work like that. Similarly, no cultural group proclaims that they offer useful falsehoods. Our prejudice is that it is the truth that will set us free.

In practice, giving priority to accuracy and accepting the costs of a more reflective rationality is rare. If interests are rooted in reproduction, the stable clusters of interests that we observe should either be directly reproduced as cultural features or should serve other reproductive interests. For most people, such as the ordinary patriotic citizen of Turania, factual accuracy only has an instrumental value in service of their primary reproductive interests. Knowing the correct price of apples at the supermarket is important, but only because that helps the patriotic citizen further her deeper purposes in life, such as the prosperity and reproductive success of her children, the status and continuity of the Turanian nation, and the propagation of those cultural items she considers most sacred.

Public support for science also usually takes this form: science is good because of applied science. Applied science gives us the power to further our primary interests by curing diseases, promoting economic development, and giving our nation more powerful weapons. It is hard to imagine any public paying the substantial costs of doing science if the only benefit they obtained was the knowledge itself, especially when resources directed to basic science would mean fewer armaments for the troops or more starving children. Most of us don't care that much about dark matter. Whenever knowledge has high possession costs and pursuing knowledge is in competition with primary reproductive interests, we should expect instances where instrumental rationality will favor false beliefs.

The alternative is to elevate knowledge production itself to a cultural goal to be reproduced. And there are, in fact, communities of scholarship in which accuracy is a primary interest. In my imagined Turanian example, international historians constitute such a community, which is insulated from the pressures that led to nationalist distortions among the historians in Turania. A properly constituted community of historians needs a large measure of institutional autonomy, allowing them to pursue historical accuracy. And since historians will have a variety of interests in life, their institutions have to be structured such that their various interests align with doing good history. Professional recognition and career advancement, for example, should reward quality work. But it is also important to cultivate an intrinsic motivation for proper scholarship and to reproduce this culture among the students who will form the next generation of scholars. Extrinsic rewards and administration-imposed metrics for productivity will encourage careerism and distortions of knowledge.[34]

All this is possible, though it is probably sustainable only for a small minority dedicated to pursuing knowledge. However, we can institute multiple communities of expertise and have them work together in a common setting—call them universities—and let them teach students not just accumulated knowledge but also skills of critical reflection and respect for knowledge. And then, we can even dream of setting up our governing institutions to be responsive to expert knowledge. Even though accurate knowledge would not be a primary value shared throughout a society, we could still obtain many of the benefits of reflective rationality.

Liberal ideologies already endorse something like this vision, especially if they resist the temptation to recognize only economic interests, thereby reducing education to the building of human capital.[35] Clearly, such an arrangement is close to my interests. I remain a secular liberal, even though I have developed some heretical tendencies.

My preferences will not appeal to everyone. Religious nationalists and other right-wing populists have come to distrust experts and higher education.[36] Given their interests, I suspect that makes sense. And even beyond right-wing populist circles, my asking for blanket trust in scientific experts would likely sound too self-serving—much like a political philosopher recommending that philosophers should be kings.

I won't, therefore, claim more than that reflective rationality is a viable ideal for some of us, even with its considerable costs. It may even be possible to institutionalize a version of reflective rationality. But it is not the only reasonable political option, much less a demand of reason. My utopia will always be someone else's dystopia, and legitimately so.

Moral Magic

There is something unsatisfying about a chaos of incompatible interests and intractable knowledge costs that reason has only a limited power to order. Maybe we need something else to help reason prevail: an overarching goal applicable to everyone, an overwhelming good we all should embrace. Some kind of spiritual enlightenment or revelation from the gods that organizes the chaos and unites our efforts. An ultimate benefit that compels us to recognize the irrationality of all actions not aligned with the greater good.

Even political rivals gesture in such a direction. Liberals are convinced they represent higher values such as justice. Religious nationalists imagine a divine mission for their nation, which they may secularize by representing their nation as a higher form of civilization. No one marches into contests with their rivals declaring their cause to be their parochial interests alone. Morality, then, may be the organizing principle we need. Striving for truth might be our moral duty.

Philosophers have defended a variety of intuitions about the nature of morality. Morality must be universal and impartial: we should treat everyone in similar situations in the same way, without regard to interests, status, or identity. Trumping parochial interests is exactly what morality is supposed to do. Moral claims should be authoritative: they must compel obedience, regardless of weak-minded waffling about whether they further our personal interests. Similarly, morality must motivate: knowing the moral course of action must prompt us to do it, rather than opting out if it seems inconvenient. There must be objective moral facts, like the facts of chemistry, that we discover rather than invent. And all of this must be tied to rationality. People who possess the same information must be able to reason their way toward the morally correct course of action. Morality must combine all the attractive aspects of economic rationality, instrumental rationality, reflective rationality, and more, imposing order on the chaos.[37]

I don't think that all these intuitions can be satisfied together. Meanings are inseparable from perception, so we naturally think that moral meanings are akin to facts. Because we are social animals, we are especially attentive to opportunities for conflict and cooperation. We continually plot social strategies, look for and construct common purposes, worry about conspiracies, and obsess about retaliation against transgressors. But neither our prosocial behavior nor our backstabbing require the existence of moral facts that have to satisfy a host of philosophical intuitions. It is tempting to think of morality in terms of weird facts that are apprehended through an occult process, but that is a mistake. Even our most intellectually sophisticated moral enterprises are devices for social negotiation rather than the discovery of facts. There are no values, just things that organisms value.[38]

This does not mean that anything goes, or that reason is merely the slave of our passions. Not every cluster of interests is stable, and not every way of life can successfully reproduce itself. For example, if we are more interested in avocados than beets, prefer beets to cabbage, and also desire cabbage over avocados, that is an incoherent set of preferences. Behavior according to such a schedule of values is ripe for exploitation: we could end up trading two beets for an avocado, a cabbage for that avocado, and receive just one beet in return for that cabbage. Such a cluster of interests does not have good reproductive prospects. A reasoning process can expose such a flaw without having to go through bitter experience, and trigger a shift in interests. The moral outlooks available to us are constrained, and rational criticism can help us configure our interests.[39] But fundamentally, we still live in a social ecology with its attendant plurality of identities and moral perceptions. On top of that, we just happen to have a cadre of philosophers and theologians running around trying to turn it into a monoculture by imposing a single regulating morality.

The theologians of the world religions have the best chance of satisfying all the intuitions about authoritative moral facts.[40] They inherit the results of a cultural evolution that led to belief in omni-capable supernatural agents concerned about human morality. Their gods create the universe for a moral purpose and order the world such that ultimately there is an overriding goal with which lesser purposes must align. And then, our religious experts also point to incentives such as heavens and hells, ensuring that instrumental rationality will lead us to the straight and narrow. They evangelize insistently, reducing the acquisition costs of religious and moral knowledge. And they promote social sanctions to ensure that the possession costs of morally correct beliefs are low and that the costs of deviance are unbearable.

The theological method of obtaining moral facts critically depends on occult ways of knowing, such as revelation. But if challenged, religious traditions can fall back on a demand for absolute trust. They can represent acceptance of revelation as a precondition of a proper social order. Absolute trust in the source and purpose of morality, after all, is the moral thing to do.

A theological conception of morality, then, can satisfy many of our most demanding intuitions about morality. It is not, however, accurate—if that matters.

When secular moral philosophers take over, they usually don't retain all intuitions about morality. For example, morality might be universal but not motivating, or not the only rational option. Some people will just favor themselves and a small circle of others they care about. They may not show equal concern for everyone while not suffering from any defects in knowledge or in reasoning ability. They may behave like assholes toward those outside their circle of concern, but they are not fools. Powerful people may have little practical reason to be decent toward the lower classes. The rich can be obnoxious, but also successful and happy.[41] None of this is an obstacle to doing some very useful moral philosophy.

Those committed to a morality of equal regard will still work to over-come class divisions. Moral reasoning does not require an overriding good that all rational people must recognize.

Even so, a more hardcore conception of morality, with all magi-cal features intact, is still attractive. Many anti-religious thinkers want a complete replacement for what theological morality offers. Since be-lievers commonly associate religious dissent with immorality, a superior secular conception of morality that preserves all our intuitions would be very welcome.[42] And since occult ways of knowing are not acceptable to most nonbelievers, moral facts must somehow derive from the material facts that are discovered by science.

The best bet for finding such facts is to look into human nature. There is a long tradition of trying to delineate the natural purposes appropriate for a human being, and equating moral behavior with the actions that best promote human flourishing.[43]

Psychologically normal humans care about things such as freedom and security; it is no great stretch to call them common human values. An immediate difficulty, however, is that such values conflict with each other, and even within themselves.[44] My political environment is full of disputes over freedoms: the freedom not to wear a mask, to carry guns, to fully express a religious way of life without being burdened by secular imposi-tions. Liberal college professors such as myself do not have a lot of sym-pathy for such concepts of liberty, preferring freedom from overbearing religions. Nonetheless, right-wing freedoms are still genuine freedoms.

It would be useful to have a way to balance conflicting freedoms against one another. This, presumably, is exactly what moral reasoning is supposed to accomplish. But there is also a danger of trying to convert a political negotiation into a technocratic calculation: solving an optimiza-tion problem of maximizing freedom or security or some ideal combina-tion. That won't work.[45] Reasoning with one another can help fashion a compromise. But all such settlements will be temporary and be shaped by local circumstances, reflecting competing collective ways of life with distinct notions of human flourishing. These compromises will depend on coercive power as much as the force of reason.

To make matters worse, human nature is not constant. In an era of rapid biotechnological advances, it is not difficult to imagine a capability to change human nature. For example, we might bioengineer men to be less aggressive. Is that a good idea? If morality proceeds out of human nature, could moral reasoning properly answer such a question? After all, altering human nature could also change our notions of human flourishing.[46]

Any practical reasoning, whether instrumental rationality or moral deliberation, requires a stable background of interests. Even the kind of personal transformation envisioned by religious traditions strains this framework, but such transformations are usually difficult and costly and are supposed to proceed according to predetermined moral ideals. But what if we had cheap and rapid means of shuffling our deepest interests and changing our basic nature? The result would be a kind of moral vertigo. By entering a brain-zap machine, we could either instantly acquire the steely resolve to confront injustice or achieve a state of bliss where we are relieved of worldly cares. What, then, would we do? With no stable interests, we could only act on our immediate desires. Our ability to reason practically about long-term interests would break down.[47]

Science cannot deliver hard moral facts, because there are no such facts. Reason cannot support a demand for factual accuracy in all circumstances.

If we happen to want accuracy, we have some methods that work well. We have scientific and scholarly enterprises that can help us achieve accuracy. Sometimes understanding the facts correctly is in the interests of all actors. Our climate crisis, for example, is increasingly threatening a collapse of civilization. In these conditions, science denial, which paralyzes our ability to respond, undermines the long-term interests of almost everyone.[48] In many a bad science-fiction plot, an alien invasion causes humanity to set aside its internecine quarrels. But in the absence of such an existential threat, our interests will diverge, and so will our interests in supporting and accepting science.

Harmful Truths

Years ago, after I had my sleep paralysis and hallucination episode, I wanted to know what had happened. I already knew a little about the underlying psychological and physiological processes; I hit the books and found out more. Slow-moving, deliberative reasoning processes led me to become even more convinced that nothing paranormal had taken place. I don't know what I would have done if, by coincidence, my wife's plane had crashed. In an alternative timeline, I could have become a believer in premonitions. I would have been mistaken.

I then completed my doctorate, and eventually found a faculty position and became comfortable as part of a scholarly enterprise. I have, I hope, made accuracy one of my main purposes. I have kept nationalism and religion at arm's length, avoiding entanglements with any community that demands conformity of belief. I care about accuracy, though I am wrong often enough to be regularly embarrassed.

My students all favor a concern for accuracy, especially when it comes to weirdness. My Weird Science students readily engage in political and moral discussion in the classroom. Occasionally, we encounter conflicts of values that we can't resolve, but the students agree that critically examining weird claims is a good idea, as long as the criticism is fair. If the claims are confirmed, what used to be a fringe notion may become exciting, revolutionary new knowledge, with immense practical value. And if the skeptics are right, that is also good to know. After all, many forms of weirdness, such as vaccine resistance, can do a lot of harm.

Some of my more enthusiastic students get carried away and propose that a course on weirdness should also be available to high school students. But then, Weird Science students are largely a self-selected group already interested in the subject. They have been taught to value education for much of their lives, and they are knee-deep in student debt to prove that they do. Still, I can count on something of a consensus: respectful criticism of weirdness is a good idea.

Outside of my classroom, my experience has been more mixed.

Once, in a history of science conference in Europe, I presented some research I had done together with my wife on some distortions of biology by conservative Muslim apologists. Some of the relics of medieval conceptions of biology and gender that had surfaced in the apologetic arguments were especially interesting.[49] After my talk, someone in the audience questioned whether my criticism of Muslim weirdness was appropriate in a climate of Islamophobia. Apparently, my work could be construed as an attack on vulnerable communities.

I have had to dance around such concerns on other occasions as well. In a volume I co-edited on Muslim apologetic distortions of the history of science, we had to start by carefully disassociating our criticisms from popular narratives of Islamic cultural inferiority.[50]

Now, I am annoyed at having to deal with this. Weirdness is universal, and even the widespread abuse of science and history for Muslim apologetic purposes does not necessarily mean much other than that Muslims are human. There is no shortage of anti-Islamic nonsense either.[51] I resent, however, suggestions that I should hold back on criticizing weirdness in an academic environment. There is already too much of an atmosphere of puritanical moral rectitude on university campuses; I'd like to have less.

My annoyance has a lot to do with my idiosyncratic interests and temperament. I don't like the puritanical impulse. Arthur C. Clarke, one of my favorite science-fiction authors, once remarked that the greatest tragedy of human history may have been the hijacking of morality by religion.[52] I think it should be the other way around: the tragedy was the hijacking of religion by morality. There are good reasons to suspect that the spread and success of world religions have been helped along by the way these religions have joined supernatural beliefs with social morality: heavens and hells, gods who surveil our every move and punish violations of moral codes. If so, we have created monsters. Supernatural belief on its own could be benign. The faithful perform some impressive rituals and regularly get together with friends for feasts, all for the price of accepting some weirdness that is harmless from most points of view. Religion as a social club can accommodate a desire for accuracy as well: old-time pagan gods are easy to convert into modern psychological metaphors, expressing how we can be moved by impulses beyond our control.[53] I am not averse to praising the parking gods upon finding a spot. It's the blasted moralists who ruin everything with their moral panics and witch hunts and thought police. And moralists can be secular as well as religious.

Having done my rant, though, I have to acknowledge that criticizing weirdness is not necessarily harmless. Some very unsubtle Islamophobes have pounced upon some of the same Muslim misrepresentations of science that have attracted my attention, and weaved legitimate criticisms into their narratives of Islam as an uncivilized plague.[54] And beyond that, any criticism of weirdness that is associated with what communities hold sacred is bound to cause offense. That doesn't bother me: I take a label of "sacred" as an invitation to poke at something with a stick. But I certainly cannot count on any consensus that such poking is harmless, not even on university campuses. I can easily imagine the sort of indignant self-parodying accusation that questioning the beliefs adopted by marginalized peoples is an unforgivable expression of privilege.

Therefore, my hopes for skepticism about weirdness are more minimalist than what the unrepresentative opinions of my Weird Science students might suggest. Within scholarly institutions, or at least in those parts of the academy that are more concerned about facts than meanings, I hope that we can let criticism run free. I would really like to see the political left get over its puritanical moralism, especially since my political inclinations are pretty leftish.[55]

Whether poking at sacred beliefs is allowed to escape beyond the sciences and similar disciplines centered on accuracy is a question for political negotiation. I would prefer that it did escape, not least because I think that science can offer all of us much more than an infrastructure for obtaining more potent drugs and more powerful bombs. But I can't with any honesty claim that thinking otherwise is immoral or irrational.

The Successful Scientist

There once was a scientist who did everything right, and was resoundingly successful.

She grew up in an upper-middle-class professional household, and her parents planned her every activity with a view to future success. She went to the best schools and got the best grades.

Her teachers found that she was good at math and very good at building things. And since these were the best schools, the teachers let her use her imagination. Her dolls built spaceships and went exploring distant planets.

She was successful in the private academy she attended and fully bought into her parents' plans to get her into a top university. And with her stellar performance, high school projects imagining the exploration of habitable planets orbiting far-away stars, and the sound financial position of her parents, she got into one of the best universities for science.

Taking the advice of her professors, she made a plan for her years as an undergraduate. Following her ambitions to explore distant planets, she would pursue astronomy. The fantastic pictures of galaxies and nebulae that adorned her childhood bedroom walls would become even more real to her.

She declared a physics major, studied hard, solved lots of equations, and found herself spending nights at the lab and the observatory. She did summer research projects and got her name on two astrophysics papers. As an undergraduate, she was resoundingly successful.

Naturally, she got into a top graduate school with a very highly ranked physics department. She then planned her graduate education. Exoplanets were romantic and exciting, but would pursuing them work as a career? There was a large overproduction of science doctorates, and getting to the top of the profession by securing an appointment as a research professor in a leading university was getting more difficult every year. Scientists were rewarded not even on the basis of papers they published but the amount of grant money they brought in. Support for exoplanet research was minimal. The real money went to applied fields. So she abandoned astrophysics and went into materials science.

Science was more difficult for women; it had a culture of workaholism that assumed scientists would have a spouse to look after all nonscientific aspects of their lives. But then, there were also some newly instituted support structures. She applied for and won a prestigious national fellowship for young women scientists.

She spent all her waking hours in the lab. She became the lead author and co-author on a string of important publications and became known as a promising hotshot developing materials to improve computer performance. In graduate school, she was resoundingly successful.

Then came a postdoctoral fellowship, with two more years in the lab trying to beat competing groups to develop materials. She got lucky and was closely associated with a breakthrough on a material to improve computer memories.

She was then well positioned to secure a faculty job at a Big Name university. There was nothing to plan after that. The next five years of her life was a race to set up her new lab, strengthen her connections with industry, and establish her lab at the cutting edge of research. She did an insane amount of work, put out a string of publications that maximized her numbers according to scientific productivity metrics, and pulled in impressive amounts of grant money. She was resoundingly successful, and was rewarded with tenure.

She could then plan again. Prestige depended on the grant money and high-impact publications associated with her lab. She needed to more strongly orient her lab toward industry and strengthen its position with regard to patents and intellectual property as well as scientific publications.

In mid-career, she found herself in a position analogous to a CEO of a large group of lesser scientists. She managed extensive collaborations and came to influence the flow of research money at the largest corporations and the most important government agencies. She became equally at home among Silicon Valley billionaires as in academic conferences. She was an excellent political infighter, directing money to her projects and to others she favored and expected favors from. Most of the daily work in her lab was now done by postdoctoral fellows, research associates, and graduate students. But she took pride that unlike some of her competitors, she still regularly got her hands dirty in her lab. And she was resoundingly successful.

Later, as she became more involved with the national and international leadership of science, her time in the lab diminished further. But she became a role model for women in science and a spokesperson for efforts to combat math anxiety among girls. In one public appearance, she went into a spontaneous rant against the fake science in Scientology; a video of her rant went viral. She started to make more media appearances and presented a popular television special denouncing anti-vaccination beliefs and conspiracy theories.

Her public and leadership roles confronted her with scientific questions of public importance, especially climate change. Knowing that retirement was not so far off, she sat down to plan again. Her intimate knowledge of both the science and the political system made it clear that very little would be done about the climate crisis. Unless some technological fix could be found—and nothing plausible was in sight—civilization would likely collapse in another few decades.

So she adjusted her investment portfolio and purchased a fourth house in a remote, easy-to-defend area with plenty of secure storage for energy and food stocks. Many of her tech entrepreneur friends had already been planning for such an eventuality. A major problem was that hired armed guards could not be trusted; they would very likely defect to the local warlord. The best course of action was to replace the human element with fully AI-controlled security systems. She invested heavily in the further development of such systems.

All these were just precautionary measures. She planned to be dead of natural causes before the worst happened.

And she was resoundingly successful.

6 REVOLTS AGAINST EXPERTISE

Truth and Threats

Toward the end of the semester, I like to have my Weird Science students explore topics that diverge from the more familiar examples of fake science and paranormal beliefs. After all, if they have learned something about the nature of science, they should be able to apply their new knowledge beyond questions about weirdness. They should even be able to examine science itself with a more critical eye.

On some occasions, my students have looked at ideas from the cutting edge of theoretical physics that sound like science fiction. Some, inspired by claims of alien visitors, have been curious about the prospects for faster-than-light space travel. Some have been captivated by the popular image of string theory, with its hidden extra dimensions of space. And then there are the multiple universes that naturally appear in many physical cosmologies. Many such speculative ideas can be virtually impossible to test even after decades of theory development. Some physicists accuse others of losing contact with reality, of not practicing proper science. Some philosophers argue that theories can be supported through nonempirical considerations.[1] Are these similar to the evasive maneuvers that some defenders of weirdness engage in? What can my students say, as outsiders, about matters where the experts disagree?

More recently, my students have gravitated toward discussing existential threats. Since the mid-twentieth century, we have lived with the possibility of a nuclear war annihilating modern civilization. My parents' generation grew up with duck-and-cover exercises in school. And now, my students grow up under the shadow of a climate crisis that might lead to a worldwide collapse. As our technologies advance further, we will be able to destroy ourselves through genetically engineered pathogens, super-powerful artificial intelligences slipping out of control, or environmental degradation resulting in a catastrophic decline in biodiversity.[2] Since such existential threats are all connected to applied science, they lead to interesting questions. Should we support science and technology, if we risk destroying everything in the process?

Normally, my students unreflectively think that technology is wonderful. If I ask them to offer some criticism, their first impulse is to think about social media abuses and cell phone addictions. Beneath the surface of their positive attitudes about technology, there is a certain resignation: technological change is something that just happens, as a force of nature. All they can do is adapt. Many of them think that cartoonishly evil tech billionaires are the best representatives of science and technology.

Existential threats expand my students' horizons, making them more willing to question our technological choices. Moreover, many existential threats are exacerbated by the sort of science denial and ignorance associated with weirdness. Most of us ignore the possibility of nuclear war, trusting in experts and national security establishments. We mostly ignore the prospect of biodiversity collapse, at most supporting minor environmental regulations while continuing with business as usual. And powerful business interests and election-winning political movements are invested in climate change denial.

Since science denial is so often politically motivated, our discussions regularly turn to politics in our post-truth times. Globally, we see right-wing populist revolts against scientific expertise, where leaders with authoritarian impulses denounce liberal elites for insufficient loyalty to nation or religion. Their followers make up alternative facts and affirm them in their media echo chambers. Since my students are in training to become experts themselves, they tend to trust professional expertise. With existential threats, they think that respect for science is a life-or-death matter. Our criticism of weirdness acquires a whole new sense of purpose.

The skeptical movement in the United States first organized in the 1970s. Back then, together with perennial favorites such as creationism and psychic powers, astrology was popular, along with UFOs and psychoanalysis. Some of the weirdness of that time has fallen out of fashion, such as biorhythms and pyramid power. In any case, Paul Kurtz, the philosopher who took the lead in organizing the skeptics and getting *The Skeptical Inquirer* started, spoke of a rising tide of irrationality that demanded a skeptical response.

The cultural upheavals of the 1960s had also affected trust in mainstream science. On university campuses, philosophers questioned the heroic image of science, and students against the Vietnam War associated science and technology with the military-industrial complex. The liberal consensus that had been dominant after the Second World War had fractured: the best and the brightest from the top universities had gotten the United States mired in Vietnam, and experts were perhaps not to be trusted as before. Materialist Western science, some critics said, was running up against its limits. Science needed to acknowledge the realities described by Eastern religious traditions and Western psychics. Though tongue-in-cheek, some war protestors attempted to levitate the Pentagon.

The skeptics did not get involved in the culture wars, confining themselves to criticizing weirdness. When I subscribed to *The Skeptical Inquirer* in the late 1980s, it was an established periodical which was controversial only because of charges that skeptics did not treat weird claims fairly, or that they had an overly narrow conception of science. Still, while not concerned about the political questions of the day, skeptics had a definite overall outlook: science was central to achieving a secular form of human progress, which should not be set back by superstition. In the 1970s, as the skeptics were organizing, Paul Kurtz even warned against "cults of unreason" tied to fascist and totalitarian politics, observing, "Today, Western democratic societies are being swept by other forms of irrationalism, often blatantly antiscientific and pseudoscientific in character."[3]

In hindsight, there was no rising tide of irrationality. Some weirdness is always popular; other varieties go in and out of fashion. Biorhythms and pyramid power have nearly vanished, and UFOs have suffered from a lapse in popularity, but conspiracy theories and ghost hunting have become more prominent. Corporate-orchestrated science denial used to obfuscate the dangers of smoking; today, it muddies the waters about how industrialized civilization changes our climate.[4] Creationism, psychic powers, and alternative medicine are still going strong. Kurtz's talk about irrationality and fascism may have had an element of professional-class status anxiety after the postwar liberal consensus fell apart.

But then again, maybe Kurtz's concerns were only premature. In the 1970s, the Religious Right was still young, and the ethnic and religious populism that has come to color politics around the world was not yet noticeable. The ruling elites of nations that had escaped colonial domination promoted secular ideals of progress. The threat of nuclear war was not yet joined by climate change, and science deniers were not attempting to kill us all.

Today, *The Skeptical Inquirer* keeps track of climate change deniers as well as psychics. Most skeptics I know, like meritocratic liberal professionals around the world, are increasingly worried about right-wing revolts against expertise.[5] Even mainstream political commentators whose main virtue is a lack of imagination worry about fascism.

Those of us who make a living from our expertise respond to the revolt against our authority by means we know best. Journalists write articles denouncing our post-truth political behavior. Academics produce books that identify political lies and reveal the fallacies of reasoning and cognitive biases that dishonest actors exploit.[6] They sound much like skeptics confronting weirdness. And professionals rally around centrist, status quo conservatism. As with the postwar liberal consensus, we want the brightest people with degrees from the best schools to be in charge.

Our predicament is more complicated than what our remedies suggest. To some extent, we experts have caused our own troubles, not the least because we have failed to recognize when expertise has been corrupted. And like most weirdness, right-wing populism is about far more than fallacies and biases. Weirdness persists, and if we misunderstand the political forces supporting it, so will right-wing science denial.

The Creationist Example

Many forms of weirdness are as ephemeral as biorhythms: their novelty wears out, they fail to get institutionalized, and they fade.

Psychic powers or ghosts have staying power. They are too firmly rooted in features of normal human minds. Such beliefs do not need to be organized. If psychic beliefs are supported only by New Agey, individualist spiritual movements, they may become popular but remain socially and politically insignificant. We can have secular societies without a majority turning into scientific skeptics.[7]

Some weird beliefs, however, are not just deeply rooted but solidly institutionalized and socially potent. Nationalist distortions of history, for example, seem inevitable. Creationism is similarly resilient. We do have a psychological tendency to perceive purpose and design everywhere, and to think of life forms as manifestations of immutable essences.[8] But most of the force of creationism is due to the institutions of conservative religion. For many decades now, creationism has been associated with right-wing populism, challenging the scientific expertise revered by professionals. Our long experience with creationism should have some lessons for our post-truth times.

The 1960s revolt against the liberal consensus was not confined to those exploring Eastern mysticism. In the United States, the revival of young-earth creationism was also gaining steam. The Religious Right would emerge from the 1960s by organizing an even greater suite of conservative cultural resentments. Cold War liberalism had responded to the Soviet lead in the space race by investing in science education and further professionalizing the U.S. educational system. Religious nationalist perceptions of reality were challenged not just by more rigorous biology instruction, but also by social science and school desegregation. The government imposed liberal expertise on a deeply conservative public, and a backlash followed.[9] Creationism became part of a right-wing milieu that combined everything from the conspiracy thinking of the John Birch Society to antifeminist activism.

What distinguishes American creationism from random crank notions or religiously motivated passive resistance to evolution is its institutions. The revived young-earth creationism was led by engineers and ministers who wanted alternatives to the secularized academic world. It was not enough to have popular literature attacking evolution; creationists needed organizations like the Institute for Creation Research and their own peer-reviewed publications and conferences. Mainstream academics observing the creationists thought of these as a bizarre echo of real scholarly institutions. Creationists simultaneously denounced secular expertise and advertised the credentials of creationists who had doctorates.

Even today, the leading anti-evolutionary institutes and think tanks are small affairs compared to evolutionary bastions such as universities. But the apologetics factories are only a small part of organized creationism. There are also churches, private religious schools, and right-wing media. Within the right-wing subculture maintained by such institutions, the trusted experts dismiss uncongenial parts of secular science and promote loyalty to a conservative Christian description of reality. Creationist leaders typically also validate other right-wing preoccupations, from traditionalist, patriarchal sexual morality to lower taxes. Creationists imagine plots to deny the evidence of creation, and then endorse conspiracy theories about the Deep State undermining right-wing politicians.[10] I go online to keep up with what intelligent design proponents are saying and end up spending the day in an alternative universe. In this right-wing world, reality is completely different from how it is perceived in my liberal university-faculty social bubble.

Connections to other right-wing beliefs reinforce the continuing plausibility of creationism. Religious institutions anchor a successfully reproducing way of life that is not easily unsettled by accusations of cognitive bias. Opposition to the demonic figure of liberal elites who aim to undermine religion, the family, and the free enterprise system further solidifies the mutual support among right-wing factions. Different currents within the right-wing environment—religious nationalists, business interests, libertarians, gun enthusiasts—end up warmly supporting one another instead of being an uneasy coalition. Creationists readily denounce environmental regulations as elite impositions on business. Forms of science denial pioneered by Cold War physicists supporting the tobacco industry get adapted by fossil fuel interests for use against climate science, ending up with conservative Christians thinking that human-caused climate change is a hoax.[11]

Deep in the right-wing social environment, the costs of acquiring accurate knowledge about biology are high, and the costs of going against religious loyalties are higher still. Denying evolution produces no disadvantage in everyday life: as with most of the conceptual frameworks of modern science, evolution has very few everyday consequences and is distant from applied science. Indeed, creationists are very supportive of applied science. After all, they often make their living in a modern, technologically sophisticated marketplace. Creationists are attracted to an elaborate fake science as a way of harmonizing their religious belief and their trust in technology. And creationists do not indiscriminately dismiss expertise. They trust many experts in business, law, applied science, and economics as much as any liberal. Instead, they target ideologically contentious areas of the sciences and humanities and their associated experts in regulatory institutions and government bureaucracies. Right-wingers want to replace the entrenched secular liberal experts with an alternative set of experts they have cultivated. Some dream of a complete institutional takeover that would re-Christianize the country.[12]

In the United States, there are few explicitly creationist experts who can plausibly take over. Religious conservatives enjoy some success in government and the judiciary. But overall, creationist institutions are weak. That, however, is just in the United States. Right-wing populism is a global phenomenon. In some countries, populists have been far more successful in displacing secular expertise.

The Turkish example is particularly illuminating. The elites who founded the Republic of Turkey in the 1920s wanted to escape Western colonial domination, and saw science and technology as the key to joining the modern world. They imposed a cultural revolution on the devoutly Muslim Turkish population, ensuring that secular expertise dominated public life. This included a full acceptance of evolution in education. During most of the twentieth century, Turkey was the leading example of secular government and westernization among Muslim countries. Today's Turkey, in contrast, represents re-Islamization and a triumphant right-wing populism. Turkey is also the center of a specifically Muslim form of creationism that enjoys public support and is visibly present in education.[13]

Resistance to evolution is very common among Muslims everywhere. But resistance usually leads to passive rejection, not the creation and promotion of a detailed fake science and alternative forms of expertise. Turkish creationism has roots in anticommunist religious brotherhoods in the 1960s, but it did not become very publicly noticeable until the aftermath of a U.S.-supported military coup in 1980. The coup eradicated left-wing political options and brought an alliance of religious conservatives, nationalists, technocratic centrists, and business interests to power. Creationism started appearing in public education, and controversies over evolution became a regular culture war item. By 2000, Turkey even had a private sector creationism that started to seek markets in other Muslim populations.

The Turkish creationists have been more successful than their American counterparts in institutionalizing their ideas. In the United States, creationism remains locked out of most university science departments, prestigious media, and the intellectual high culture. In Turkey, where right-wing populism has become politically dominant, religious conservatives influence much of the media and are a large presence in intellectual life. The credentialed experts who appear on television programs are often Islamists or moderate conservatives with government loyalties.

Turkey has long had a religious education system along with ordinary public education. Today, the government favors the religious schools and the private schools run by religious brotherhoods. It is not unusual to encounter university faculty who reject evolution, not just in the applied sciences but also in biology and science education. University administrations and the faculty in the newer provincial universities are often right-wing loyalists. Not just creationism but also traditional medicine derives prestige from being perceived as culturally more authentic. Nationalist falsifications of history are everywhere.

In Turkey, then, an alternative set of right-wing institutions and their favored experts have partly replaced secular experts. The religious conservatives only partially reject expert knowledge; they embrace the expertise associated with business, economic development, and technology. Islamists have long adopted modern finance and global business practices, and have been much praised in the West for their free-market dynamism. There is still a wealthy secular liberal elite, who still hold the older and more prestigious universities. Liberal political parties, however, cannot win national elections. In any case, beyond culture war items, the liberal economic and political vision for Turkey is not too different from that of the ruling Islamists.

Some other countries with right-wing populists in power have similarly institutionalized fake science. India, for example, was also led by a secular elite after independence. Even the Indian constitution promotes a scientific temper. Today, India has come to be dominated by Hindu nationalism. In areas such as economics and large development projects, almost all Indian politicians respect the expertise of the financiers, the

International Monetary Fund (IMF), and the World Bank. Indian applied scientists have found success in the global economy. But schools now teach nationalist myths about history. Hindu resistance to evolution has not inspired much of a fake scientific response, but quantum mysticism, Indian astrology, and modern variants of traditional Indian medicine flourish. Alternative experts with right-wing political loyalties have become entrenched in Indian scholarly institutions.[14]

In all such cases, right-wing populists notionally embrace science. They desire, however, a more pious, more culturally authentic form of modernity. They want the wealth and power that accompanies advanced technology, including the social technologies of corporations and financial markets. But they reject the secularity and the corruption of traditional forms of morality that have accompanied modernization. Aspiring, culturally conservative factions among elites mobilize political constituencies deprived of cultural capital and left behind by a secular liberal order.

Movements to achieve a pious modernity and forms of fake science such as creationism mutually support one another. Right-wing populism nurtures a social environment where some forms of weirdness are plausible and belief is advantageous. When weirdness is closely associated with political and religious movements, skeptics and scientists face a very different challenge compared to beliefs in biorhythms or psychic powers.

Fighting the Right

Skeptics and scientists have been fighting creationism for decades. Creationism, therefore, is also a test of ways to limit politically connected weirdness. Our history can help us figure out what works.

There is a boatload of anti-creationist literature, from short articles to hefty books. Critics have made long lists of creationist fallacies and cognitive biases. They have enumerated how creationist claims violate the scientific method. More valuably, they have addressed the specific

scientific mistakes creationists make. The best argument against creationism is to show how evolution provides a superior explanation of
biological evidence. That job has been done.

All this effort, however, does little to persuade anyone who does not
already trust scientific expertise. Believers immersed in a right-wing
milieu encounter creationist apologetics. And creationist principles for
discerning truth from falsehood include testing claims against the testimony of scripture. Apologists elevate common sense and conservative
moral intuitions against attempts to blind the faithful with science. And
they associate evolution with a liberal elite that does not deserve trust.[15]

The first instinct of skeptics and scientists, then, will be to prepare
the ground for trust through better science education. Practically everyone in modern societies goes through secondary education, which includes basic science. A sizeable minority go on to college. We have a large
professional class whose social and economic status depends on their
education and credentials. Scientists and skeptics are mostly part of this
professional class. Therefore, we naturally associate weird beliefs with a
failure of science education. Those of my Weird Science students who are
convinced that weirdness is harmful instinctively reach for education as
an antidote.

Education is certainly important for promoting trust in the experts.
It is less reliable for imparting any deep understanding of science.

I sometimes teach physics to first-year nonscience majors. They are
a bright, talented lot, but high school makes them think that science
consists of a set of facts to collect like stamps. Worse, they dislike math.
I teach the course with a science-fiction theme, doing lots of relativity,
black holes, quantum physics, and cosmology. We read and discuss a
hard science-fiction novel that incorporates a good deal of physics. In
the end, though, I hope that they will become better readers of science
fiction and popular science books. I cannot expect them to develop any
serious understanding of quantum mechanics. At best, I might install my
voice of authority ringing in their heads, reminding them to be cautious
when they encounter bullshit claims of quantum magic.

The basic conceptual frameworks of today's science, such as quantum physics or evolution, are counterintuitive. They require considerable intellectual and mathematical maturity to understand. What I can hope for in a general science education course is to help students appreciate science and have a better-informed trust in scientific expertise.[16] I love teaching nonscientists, but the knowledge they acquire is inevitably shallow. And if their sole motivation to be in my classroom is to fulfill a requirement for a credential, they will find a way to pass the course and then rapidly forget almost everything I say.

Does science education reduce beliefs such as creationism? Somewhat.[17] Among my first-year nonscience students, a few will occasionally answer an exam question correctly, but add a note that they cannot accept big bang cosmology. If their religious loyalties have them reject cosmology, my course will not change their minds. The third- and fourth-year students in Weird Science know that evolution is no minor detail in science, and they try hard to reconcile it with their spiritual convictions. Their education often inspires a more liberal religious compromise. I occasionally ask my colleagues in the biology department; they estimate that perhaps as many as a fifth of their biology majors graduate with doubts about evolution. Premedical students are particularly likely to remain creationists.

Some biologists argue that science education should more directly confront creationism, and should even question the religious beliefs that motivate rejection of evolution.[18] But even then, it is not clear that science education can affect much. Recent polls show a small drop in creationism in the United States, but this is probably due to the overall decline in American religiosity.[19]

Insistence on more evolution in public education can, however, inspire a more organized creationist backlash. Social conservatives have different ideas about education than most scientists. Conservatives often conceive of knowledge as less of an expert domain, preferring it to be shaped by the practical needs of communities. Education should instill the appropriate loyalties, respecting the organically developed, morality-infused perceptions of reality nurtured by religious traditions. If a liberal education undermines these loyalties, then it should be made less liberal.[20]

Evolution, then, is caught up in a political struggle over the means of cultural reproduction. Skeptics and scientists like to think of science as nonpolitical objective knowledge. But in practice, science gets associated with the liberal side in a political rivalry. Our institutions of science education are closely tied to a liberal, meritocratic concept of education. Those of us in science and higher education—people like me—usually help to reproduce liberal attitudes. We stand against not just conservative religiosity but also any bonds of loyalty and solidarity. Instead, we trust in rules, processes, and bureaucracy. We inject competition, including market competition, into all public relationships. And since the professional class whom we train and are a part of has expanded and come to dominate the upper incomes and political power, we are happy with the status quo. We think that rewards should be distributed according to merit, without regard to loyalties—doing otherwise would be corruption. People merit their station in life, except where markets and bureaucracies are distorted by prejudices of gender or race. And merit is made manifest not just in economic rewards but also in prestigious educational credentials that signify the possession of high-quality human capital.[21]

Creationists, whom liberals think of as ignorant peasants, regularly appeal to traditional values and community control, trusting that democracy will reward populists. To beat back the rabble, meritocratic liberals favor top-down, elite interventions. In the United States, creationism has only been kept out of public education through court decisions that have ruled that teaching creationism is unconstitutional. And to strengthen evolution in science education, liberal experts rely not on democratic persuasion but nationally imposed standards. In Turkey, secularists have attempted to ban Islamist political parties that supported creationism by trying to have the courts declare them in violation of constitutional secularism.

Right-wing populists frame all such conflicts as liberal elites obstructing the will of the people. Beyond the culture wars, however, the real contest is between rival elite factions. U.S. courts, for example, might seem sharply divided between liberals and reactionaries, but most legal disputes concern business, and jurists of all political allegiances are in concord about subservience to business interests. Professionals usually favor the status quo, but right-wing elites leverage popular cultural resentments and distaste of bureaucracy to gain an advantage. Mostly, however, business as usual prevails.[22]

On balance, the politics of professionals is friendly toward science. Professionals tend to be more secular, more liberal, more confident that knowledge will deliver human progress. Science has provided citizens of modern states with the means to intensify our plunder and increase our wealth, giving us the power to realize our dreams of freedom. Traditional loyalties, liberals think, lead to bondage and backwardness instead. But religion and ethnic nationalism still persist. Right-wing populists also love technology, and they can plausibly offer a promise of a pious modernity. They can also deliver plunder and wealth, and hint at freedom from bureaucrats and petty managers. Secular liberals, it seems, are now behind the times.[23]

Education is difficult. The culture wars are a mess. Skeptics and scientists have few tools to counter creationism, and they don't seem to work too well. We are caught in a political contest where we are out of our depth, and we have little influence over events.

Competitive Advantages

Skeptics and scientists prefer a politically neutral, just-the-facts approach to weirdness. We stick to the science and point out the psychology that makes us susceptible to false beliefs. Trying to understand right-wing populism draws us into less comfortable social and political ways of thinking. Still, if we had a good explanation of the appeal of right-wing movements, we might be better able to address politicized forms of weirdness.

Plenty of journalists and political commentators have analyzed our post-truth environment, explaining right-wing success through heightened racism and xenophobia or the dastardly machinations of Russian agents. Even if true, however, such explanations would be superficial, like blaming Balkan and Middle Eastern wars on ancient hatreds. Why is ethnic and religious nationalism so compelling *now*, worldwide? The outsize influence attributed to Russian meddling strikes me as conspiracy thinking; commentators who deploy the rhetoric of reason and truth have themselves gone down the rabbit hole.[24]

I also have to confess a prejudice against mainstream American journalism. I come from the Middle East, and I am not happy with how journalists treat the American empire as the benign, natural order of things. I cannot forget the lies amplified by the press in the run-up to the Iraq War and other episodes of destruction visited on the Middle East. Nor am I inclined to forgive media figures for their peculiar sense of balance that long made climate change denial seem respectable. American journalism has been famous for its conformity, groupthink, and post-truth reporting long before the term was invented. Journalists rarely see beyond the narrow spectrum of disagreement between our two conservative major parties. I don't trust them. Much of their reporting about our post-truth times has the tone of commiseration within the professional class of how the lower orders don't listen to their betters anymore.[25]

Hand-wringing over deplorables happens in the academic literature as well, even leading to distorted studies.[26] Nonetheless, right-wing populism has been the subject of intense historical and social scientific scrutiny. I am in no position to evaluate the literature in detail; this is a case where I have to defer to the experts.

There does not appear to be a scholarly consensus about the exact causes driving right-wing movements. Still, there are some common themes. The full-spectrum dominance by business in the last few decades has coincided with a global rise in right-wing populism. Many scholars suggest that this is not an accident. Privatizing everything and inserting markets and competition everywhere appears to have created conditions

favorable to right-wing movements. Increased market discipline promotes inequality and insecurity. And insecurity reliably correlates with
intensified religious and paranormal belief. Insecurity also goes together
with a backlash against a liberal loosening of social morality and the increased self-assertion of women and minorities.[27]

The politics of educated professionals centers on meritocracy, economics as the measure of all policy, and bureaucratically enforced recognition of disadvantaged identity groups. Professionals do not challenge
the rule of bosses and landlords, particularly since many of the economic winners of today are workaholic professionals with the best human
capital certified by expensive degrees. But while the natural ideology
for professionals has become meritocratic liberalism, strictly individual,
merit-based competition in markets defined by rules is not a stable condition. Ruling out economic actors that engage in fraud or collaborate
in a mafia-style of free enterprise requires heavily bureaucratic micro-
management. And none of that prevents powerful actors cooperating to
rewrite the rules in their own favor.

It turns out that right-wing populism enjoys a competitive advantage. Members of the lower strata of societies, who end up managed by
professionals and locked out of economic advancement, resent their immediate overseers. Many identify their problems with the bureaucracies
so loved by liberals, and their fake diversity and inclusion. While meritocratic liberals have suppressed forms of solidarity such as labor unions,
religious and ethnic loyalties remain. Some elite factions can improve
their standing by mobilizing such loyalties among the insecure middle
and lower classes, affirming markets as the arbiters of value but inter-
preting the liberal protection of marginalized identities as unfair competition.[28]

Right-wingers in different countries exploit residual solidarities in
different ways. In Turkey, for example, the main private-sector promoter
of creationism was a cult-like organization that owed as much to organized crime as more traditional religious brotherhoods. Modern Islamic
movements that use religious loyalties and mutual aid to gain economic

advantages have been critical for Islamist success in Turkish politics. Some movements have even specialized in recruiting pious and talented students and supporting their training in professional fields and the applied sciences, in order to form an alternative elite to replace secular liberals. Factory workers and bosses pray together, but the devout management uses up-to-date and efficient business practices to minimize labor costs.[29]

The friends I keep in touch with in Turkey are almost all highly educated secular professionals. Their children go to expensive private schools that try to avoid creationism. They vote for liberal centrist politicians. When we get together, my friends invariably complain about the authoritarian tendencies of the long-ruling Islamists, worrying about pious interference with secular lifestyles. My friends are meritocrats: they highly resent how religious loyalties have become an advantage in obtaining positions and winning contracts. Religious brotherhoods, they think, have too much influence even in the modern business sectors. And the government practically runs on such corruption. Our conversation sometimes takes on an antidemocratic coloration. The devout, uncultured poor, some of my friends insist, are unproductive burdens on everyone else, exchanging votes for Islamists with government-organized aid that keeps them from destitution. If such people are allowed to vote, demagogues will naturally come to power.[30]

So I think it's plausible that right-wing populists take advantage of current political conditions. Meritocrats are committed to an ideal of perfect individual competition. But populists mobilize both wide but shallow commitments to religion and ethnicity and the deeper loyalties of tight faith groups. They cooperate in order to outcompete the meritocrats. As a physicist, I am drawn to symmetry-breaking as an analogy, but biological metaphors about cooperation work even better. The liberal ideal, a monoculture of perfect individual competition, fractures into competing groups that cooperate within themselves.

While plausible, accounts that rely on insecurity as a motivation for cooperation are also very general. There may also be something specific to our post-truth moment. Another important factor I have run into in scholarship about right-wing populism is the internet.[31] Online communication costs are very low, and information has not yet come under the control of elites. The result is a perfect environment for conspiracy theories as well as cat videos. Conspiracy theories that emphasize identifying enemies and trustworthy actors can be useful in coalition-building. Technological optimists had once anticipated that the internet would make low-cost knowledge available to all, helping provide a level playing field for perfect individual competition. But if lies and conspiracy theories swamp our information environment, or worse, if they help form right-wing coalitions and provide them a competitive advantage, there is less room for such optimism. At any rate, the internet is now organized as a true marketplace of ideas, drowning in advertisements and low-quality information. A market is not a good institution to promote accuracy of claims.

Secular liberals still hope to institute a global monoculture. So far, however, we have found ourselves as just one competing political ideology among others. Even our meritocratic ideals are deformed, on one side by our subservience to bosses and landlords, and on the other by our need to assemble coalitions of our own. We end up with a postmodern version of patronage politics for ethnic groups, glorified as supporting the marginalized. Liberal meritocracy, I have come to suspect, fuels right-wing reactions as much as it fights them.

Making Progress

Meritocratic liberalism is suspicious of politics. It aims to turn elections into job interviews where voters act like a collective human resources department, evaluating candidates on the basis of qualifications and credentials. It is an ideology of standing aloof from ideology, measuring performances by objective metrics, bypassing messy public deliberation in favor of technical problem-solving.

Since right-wing populists revolt only against some experts, their favored forms of governance and economic management are not radically different from the meritocrats. They are, perhaps, a bit more enthusiastic about plunder, a bit more open to cutting bureaucratic corners with a more mafia-like style of operations. A habit of disregarding science, however, has consequences. Liberal governments do not manage pandemics well, but right-wingers completely botch the job. Liberals are complacent in the face of environmental collapse, but right-wingers positively invite the apocalypse.[32] In that case, a politics of rallying behind expertise becomes more attractive. "Trust the science" is not the most exciting slogan, but scientists and skeptics can at least hope for more science-based policies.

Science nerds, however, are a small constituency. In practice, liberal politics aims to shore up the authority of all professionals. This includes lawyers, economists, and management consultants, who often indulge in predatory behavior and promote self-serving versions of expertise.[33] If such prominent professionals provoke distrust, is it feasible for scientists to claim that they are an entirely different breed of professional?

Perhaps the tech sector can help. After all, they depend on science. And most people appear to be in love with their devices. But then, the big tech companies have also acquired a reputation for addictive social media, massive surveillance, and cozy relationships with ruling elites. We have ended up with very powerful and manipulative platform monopolies. Tech leaders can still capture the public imagination with grand schemes such as colonizing outer space, but these are more harebrained public relations stunts than real science. The tendency of applied scientists to be more politically conservative and positive toward weirdness applies to high tech as well: Silicon Valley is notorious for its New Agey enthusiasms and plutocratic libertarianism.[34]

Historians of technology, such as my wife, like to emphasize how technology is shaped by historical accidents and political choices as much as technical imperatives. The ideology of Silicon Valley, in contrast, represents technology as an irresistible force directed only by a few genius billionaire disruptors, to which the peasantry must adapt. Every problem calls for an engineering solution, which the free market will inevitably supply.[35]

I therefore do not want tech culture to encroach on basic science. Science depends on shared knowledge as a public good. Silicon Valley would privatize and chop it all up into proprietary parcels of intellectual property.

No doubt rule by the tech titans would be better than rule by creationists or anti-vaxxers. We might even harness tech to fight right-wing populism, instituting regimes of censorship on social media to block fake news and conspiracy-mongering. But then, the tech monopolies are closely entwined with our governments and militaries, including countries where right-wingers are in power. I expect any crackdown on dissent from expert consensus would not affect weirdness so much as independent journalists and critics of the American empire.[36]

I worry, then, that a blanket affirmation of expertise and a technocratic politics driven by the interests of affluent professionals might not be good for science. The interests of professionals are not unified, and science has too small a constituency to matter.

But then again, maybe I worry too much. There is, after all, a unifying vision for secular liberals of all kinds: progress. Progress is the ideal of economists and the goal of tech titans. Human ingenuity is boundless; we can reach for ever-more fabulous wealth, more spectacular toys, and health that may one day extend to curing death. And much of the public respect accorded science, I suspect, is due to the promise that advancing knowledge will drive progress toward such common ends.

There is no doubt that modern times have improved the lives of vast numbers of people, making many of us more prosperous and allowing us to enjoy longer and healthier lives. It is plausible that a more scientific and skeptical culture has helped usher in such improvements.[37] Some historians may point out that our ideologies of liberty and property have been closely entangled with slavery and mass murder.[38] But for a true devotee of progress, these are just statistical blips. In the bigger picture, liberals might argue, the bloodbaths and the exploitation get averaged out, and we are left with a glorious vista of human advancement.

Many pundits continue to make such arguments: we are living in the best of times in the best of all possible worlds. This is true not just for wealthy Westerners but most of the world. The technocrats have been beneficent overlords; the status quo has delivered progress. If any nabobs of negativity should raise an eyebrow, here are some graphs. Defenders of the status quo include high-profile scientists, most notably the psychologist Steven Pinker.[39]

Pinker makes a persuasive case that civilization has reduced human violence, particularly in modern times. Trading has replaced raiding. As modern states have established their monopoly on violence, they have also pacified the societies within their borders. Even with industrial-scale slaughter on modern battlefields, today, we have a statistically better chance of living a long life undisturbed by lethal violence.

Technocratically managed market economies, Pinker argues, have made impressive progress in rescuing large populations from abject poverty. A small army of economists have been claiming this success for many years now.

Pinker and similar writers go on to claim that existential threats are exaggerated, or are technical problems with engineering solutions. They work through a laundry list of items demonstrating that human progress is real. Critics of the status quo, it would seem, suffer from a defect in rationality that causes them to accentuate the negative. Instead, we should continue to embrace science and reason, as made manifest through meritocratic liberalism. We should resist the false gods of religion and populism, lest we descend into a new dark age.

Now, I do physics and weirdness. I'm not an anthropologist, and I don't study international development or existential threats. But then, neither do most apologists for progress. The actual experts draw a far more complicated picture.

It is plausible that we live in relatively peaceful times. But what are we comparing today with? The chaos at the collapse of a feudal order? The constant warfare of agrarian empires? Anthropologists might suggest foraging bands instead. After all, this was the common human condition for most of the history of our species. I used to think that early human bands were packed with aggressive males who regularly conducted homicidal raids on neighboring bands, and I went along with Pinker's picture of primordial violence. But according to some anthropologists, the evidence for such violence in preagricultural societies is very thin. Indeed, settling down for agriculture looks like the biggest mistake in human history. Members of foraging bands who survived infant death lived longer, less disease-ridden, saner, and less murderous lives than civilized peoples. Our species was adapted to foraging, which has a better claim to reflect a common human nature than the self-captivity we call civilization. Our progress of the last few centuries consists of slapping high-tech Band-Aids onto deep wounds.[40] I love my *Doctor Who* videos and opportunities to puzzle about dark matter, but perhaps I should not celebrate civilization as much.

Claims about declines in abject poverty are also plausible on the surface. Private enterprise generates massive amounts of wealth, even if it has a hard time distinguishing between productivity and plunder. It also creates growing inequalities of wealth and power due to the ability of the already wealthy to get higher rates of return on their investments and to rewrite rules in their favor. But then again, some prosperity may well trickle down to the lower classes.

The statistics concerning global poverty may not be entirely trustworthy, due to politically motivated shifts in definitions and data. Regardless of such concerns, according to some experts, there is less to the decline in abject poverty than advertised. Recent reductions in poverty are almost entirely due to the rapid industrialization of China, converting much of its peasantry into waged factory workers. A rise in income, moreover, is not meaningful on its own. Vast populations around the world have been brought into the cash economy through dispossession and pauperization. Tribal people in India who have been uprooted by huge development projects go from living off the land and needing no money to casual labor sorting through toxic trash for a few dollars a day. Even their poverty wages show up as an increase in income in the statistics.[41]

Those of us in a comfortable position rarely appreciate how forcing foragers and tribal people into the modern economy makes them less well off. Liberals make diversity-and-inclusion noises about marginalized indigenous peoples. Very often, indigenous struggles are to *avoid* being absorbed into the global liberal order.[42]

Our reduction of poverty is also an environmental disaster. Perversely, apologists for progress point to environmental improvements in wealthy countries as illustrations of problems solved within the status quo. That improvement comes through outsourcing the pollution, together with the production, to countries with cheap labor. Meanwhile, the postindustrial professional class increasingly depends on financial services and intellectual-property rents. We make useless gestures of recycling our plastic and congratulate ourselves for our environmental virtue.[43]

In the end, I am most annoyed by how the rhetoric of progress downplays approaching environmental collapse and other existential threats. Invariably, any threat becomes a technical problem that demands an engineering solution that can be implemented within the status quo. Many environmental advocates have been sucked into claiming that a clean-energy transition that will maintain our consumer way of life is just around the corner, that with the maturing of solar and wind-based energy production the hoped-for technological fix is almost here, and that market forces and billionaire-funded foundations will save us if only we can break the political power of fossil fuel interests. They are far too optimistic.[44]

Even if we were to dodge one bullet, the status quo keeps generating new threats. When I was young we had nuclear war, overpopulation, and a general uneasiness about pollution. All three remain. Nuclear weapons continue to proliferate. If we have not had any accidental nuclear wars so far, this has mostly been due to luck. Population growth is slowing down, although that is too little too late. Concerns about pollution have expanded into our climate crisis and associated calamities such as ocean acidification. Biodiversity collapse is well on its way. Our politics is in a state of paralysis, and since we cannot change our social trajectory, we have to hope for a technological fix. If clean energy turns out to be overly optimistic, maybe we can do geoengineering. But even if such measures work, they would be another set of high-tech Band-Aids.

Meanwhile, it is easy to notice new existential threats on the horizon: super-artificial intelligences that wipe us out, bioweapons genetically engineered with ever-cheaper technologies, the insect apocalypse. A highly technological, global civilization constantly generates threats at a global scale. Can we keep riding our luck? Forever counting on human ingenuity to produce the right technological fixes is an act of faith.

I have my doubts about progress as a device to unify professionals. It's a nice myth, but like all myths, it is best not to take progress too literally. Professional interests are divided, and our current political fault lines do not neatly separate scientific heroes from superstitious villains. The professional class is not an embodiment of reason, and professionals like me are no less responsible for our predicament than the mindless tribalism we profess to oppose.

But Is It Good for the Scientists?

I should repeat: I do physics and weirdness. I'm not an expert in political analysis, even the politics relevant to weirdness. When a student asks me what I think about a political question, I tell them, but I immediately qualify my statements by pointing out uncertainties and noting that there are people whom I respect who disagree. The experts are divided.

To further complicate matters, political analyses inescapably weave personal interests together with the facts. Like almost everyone, I react negatively to political disagreement, and I am tempted to cherry-pick the experts I favor and go live in a bubble. So I tell my students that a good intellectual habit is to be a critic of one's own views, to try to see one's own tribe from the outside.

My tribe, such as it is, must be the skeptics, the nonbelievers, those who raise an eyebrow at weirdness. I don't just dispassionately note the weirdness in the paranormal websites I visit; I throw my hands in the air and rant at anyone in my vicinity. I eagerly await my copy of *The Skeptical Inquirer*, read it from cover to cover, and agree with most of what I read.

The rhetoric of reason and progress comes naturally to skeptics. I am no exception. So it was no surprise when Steven Pinker appeared on the cover of *The Skeptical Inquirer*, with an article promoting his status quo optimism.[45] Praise of a heroically imagined science and Enlightenment hopes of progress is exactly what we skeptics like to hear. And precisely because of that, we have to be especially careful. I don't think this optimism is supportable in anything close to its pure form.

Celebrating science without questioning the status quo—including how the institutions of science participate in the status quo—has become dangerous. Professional-class politics has grudgingly accepted the reality of climate change. But this has led to no end of white papers, committee reports, studies, commissions, and blue-ribbon recommendations without much actual change. Our bureaucratic apparatus has churned out a series of toothless half-measures such as the Paris Agreement, and then watched them be ignored.

Climate modelers often compare possible interventions with "business as usual" scenarios. Business as usual eventually leads to catastrophe. And our meritocratic liberals, political pragmatists, and apostles of reasonable compromise represent business as usual. The political vision of professionals—my larger tribe—has been far more entrenched in power than right-wing populist challengers. The climate crisis is our failure. If our collective paralysis in the face of a warming climate is irrational, it is our irrationality as much as those who think it's all a Chinese hoax.[46]

Some time ago, I was involved with climate modeling research as an academic side-hustle. I also support the Union of Concerned Scientists (UCS), which does some very good work. About a dozen years ago, the UCS got in touch with me, saying that they were organizing members who did climate science for a lobbying trip. There was some cap-and-trade legislation being considered that could have reduced emissions. Most of the people I talked to acknowledged that the bill was wholly inadequate, watered down, and full of magical thinking about market forces so as not to offend the prevailing superstitions. But if it passed, it could conceivably serve as a starting point for improved bills to follow.

So the UCS flew me and a bunch of other scientists out to Washington, DC. We started with some training for our roles in lobbying. The UCS people told us that we were not likely to meet any members of Congress, who would be too busy fundraising for reelection. But we would meet congressional staff, who sometimes had a surprising amount of influence. After training, we started visiting the offices of representatives and senators in teams of three: a UCS person, an economist to assure staffers that on balance the bill would not hurt the economy, and a scientist just in case there was a question about scientific matters.

Very little science came up. When I had a chance to speak, I could do little but try to impress the urgency of the situation on the staffers, summoning up all of the cultural authority of science that I could muster. It was all an exercise in futility. I must have let my frustration show after a while, so some of the staffers explained the constraints they worked under. They were all terrified, it seemed, of a twenty-second attack ad an opponent could put out about job-killing environmental regulations. In some offices, it was like talking to a representative of the chamber of commerce. They politely heard us out, but we all knew that a bunch of science types had very little influence. The careers of politicians depended on the immediate economic interests of bosses and landlords back in their districts.

My limited experience with working within the system, then, has brought me to expect paralysis. For decades now, those of us knowledgeable about climate science have been in a state of escalating panic about our prospects as a civilization. But our political and economic system has been incapable of responding. If anything, climate scientists have underestimated the dangers, being wary of being labeled alarmists and losing funding.[47] And yet, we are stuck in a rut where everyone acts rationally according to their short-term economic interests, and the result is collective insanity. An invisible hand made out of omniscient psychopaths turns out to lay waste to the world.

I am useless as a lobbyist, and few politicians I actually like—as opposed to those I consider a lesser evil—will ever win an election. But even when I stick to reading philosophy books and talking to physics students, I can't fully escape politics. Thinking about the institutional pathologies that nourish weirdness, I can't help but ask questions about the institutions of science. And I suspect that not all is well.

Superficially, science is in excellent shape. It is a vast, reasonably well-funded enterprise producing a flood of publications in every sub-subspecialty. Basic science does not attract as much money as applied research that produces intellectual property, but it escapes the austerity imposed on fields such as philosophy. Those such as myself who are fortunate enough to have a tenured position are comfortable, secure, and have the freedom to spend time on projects just because of intellectual curiosity.

Universities are old-fashioned institutions that change slowly. Basic science still mostly operates as a public enterprise producing knowledge as a public good. Our results are largely open and available to anyone who cares to use or criticize our work. Scientists compete for funding and fame, but all our work is a collaborative effort to improve the knowledge we share in common. Science is supposed to be self-correcting, and it often is, but we cannot replicate every experiment and check every calculation. When scientists pursue their work because of intrinsic rewards—satisfying intellectual curiosity—we can more easily maintain the integrity of science. We still make mistakes, but mostly honest mistakes. Science critically depends on an environment of trust. Without trust, we cannot collaborate to construct knowledge.

None of this is in a healthy state today. Neither of our current political tendencies, right-wing populism or meritocratic liberalism, properly recognize public goods. And science, like all goods that have been public, is now subjected to privatization. It is managed by bureaucrats who aim to deliver return on investment.

The educational mission of universities has slowly been decaying. For most students, college is just a necessity on the way to acquiring professional status; the more prestigious the school, the higher the rank in the meritocracy. A genuine liberal arts education is still available, but increasingly limited to a wealthy elite. For the majority, the future looks like credential-hunting by jumping through online hoops, supervised by exploited adjunct faculty who already do most of the teaching and are treated like slave labor, only more disposable. Even the sort of science education that instills appreciation for science is liable to become a luxury.[48]

Our current research model is geared toward attracting grant money and producing intellectual property for private interests. Especially in applied fields, the ethos of free exchange and criticism is giving way to the more proprietary view of knowledge that characterizes weapons research and corporate development. And with administrators imposing metrics and efficiency standards and demanding deliverables, perverse incentives come to the fore. Researchers become workers overseen by managers, and respond by gaming the system.[49]

So far, the rot has mainly affected some applied sciences where untrustworthy results and even fraud have become disturbingly common. Still, the trend is to rely on extrinsic monetary incentives and management imperatives. I can't assume that basic science will remain immune.

The main danger for scientific institutions is a loss of independence, eroding their ability to respond to internally generated intellectual questions. Barring a collapse of civilization, science should survive. In the past, nationalism has distorted history, but it has also helped spur science through a desire for national power and prestige. Soviet science has become a cautionary tale for ideologically motivated interference in biology, but Soviet physics was world-class. American science has often been racist, but that is not a permanent condition. Science is resilient; it can function in many different political contexts, selling its services to those who seek power. Nonetheless, a natural science that survives as infrastructure for technology in institutions devoted to intellectual-property production will be a diminished science.[50]

When I look at the proceedings of the international creationist conferences recently organized by provincial universities in Turkey, I am both fascinated and appalled. The faculty who speak include engineers, biologists, and theologians, all producing appealingly weird distortions of science.[51] They must partly owe their positions to their support of Islamist politics. I don't know, but I would not be surprised if many of them were connected to religious brotherhoods. But I can still recognize them as fellow academics. They are intellectual entrepreneurs responding to the incentives set by a right-wing administration.

Science certainly fares better under a meritocratic regime. And yet, I can only hope that meritocracy is not the best of all possible worlds. Meritocrats still rely on quasi-religious myths like progress. They represent the business as usual that creates environmental degradation and then is paralyzed in responding to it. They pursue technological fixes to social problems and reduce science to an appendage of their administrative utopias. If there is no alternative, we are lost.

Lies, All Lies

I don't expect our sciences to figure out everything. We might not have the resources. We may fail to ask the right questions. We might have bad luck. Or we may blow ourselves up before we discover more.

I am not so sure, however, that there are facts that are intrinsically beyond the reach of scientific investigation. There are no end of signals that we cannot detect in person, from X-rays to gravitational waves. But we continually make new instruments that extend our reach as a scientific community. There are many objects—electrons, black holes—that we can probably never grasp as intuitively as a tree or a table. But we do the math. We devise metaphors to help us understand strange objects, and then we extend our metaphors into realms far from our everyday intuitions about reality.[52]

Since I don't think that science is nailed down by any predetermined method, I don't think science has permanent blind spots. There is always a chance that we will collectively learn and adapt and do things differently, although no one can guarantee that we will take the chance. Scientific knowledge, I think, is indefinitely extensible. We continually criticize and revise and rebuild our networks of knowledge. And the networks grow. Progress in politics is a questionable beast, but science advances.

Surveying the explosive growth of science over the last few centuries, it seems fair to conclude that our methods of science and the structure of our institutions promote the advancement of knowledge. Much of this has to be due to positive feedback loops. When scientists get into a loop where experiments correct theories and our theories help us do better experiments, we converge onto results that make us very confident that science is getting a handle on reality. Moreover, achieving knowledge puts scientists in a good position to further advance knowledge. Improved knowledge leads to better instruments, which reveals new facts and further puzzles to solve, prompting more work to improve our explanations, and so on. Advancing science drives even more progress in science.

Not every human enterprise is so fortunate. Our conceptions of a more comprehensive social progress have become tightly linked to our hopes for economic growth. And in the short term, growth can also benefit from positive feedback loops. We use technology to gain wealth, and then use some of our gains to reinvest in improved technology, which then helps us extract even more wealth. There are, however, complications. If our civilization were a small-scale affair, we could ignore the waste products of industry. We could act similarly to a small tribe that burns a patch of forest, grows crops for a few years, and then moves on to burn another patch down while the forest regenerates. In a global civilization, we cannot dump our waste and move to a new planet. We start seeing negative feedbacks: our plunder makes the environment worse, which then starts making the plunder more difficult. We run into physical limits, such as with my students' calculation, where they find that the volume that civilization occupies starts growing faster than light in a few thousand years. And that calculation is wildly optimistic; it totally ignores the thermodynamics we need to account for waste. We face negative feedback due to our environmental crisis now.[53]

Since the institutions of science are supported by economic growth, and since scientists depend on increasingly high-tech equipment, even science can be affected by such negative feedbacks. And when growth stalls, our hopes for social progress will be in deeper trouble still. In some ways, social progress has been cheap. Any political thinker who identifies with our current set of elites will be tempted to announce that moral progress is real and that history has culminated in the present social order. And with growth, they can proclaim that everyone benefits other than some dissidents and malcontents. Beyond cheap propaganda, however, our different ways of life and different hopes for the future reassert themselves. For a religious conservative, progress toward my social ideals is often a degeneration, and legitimately so. It would be absurd for me to try to argue otherwise by piling on statistics and graphs. From a long-term perspective, any sense of social progress risks being washed out by shifting sets of competing interests, the vulnerabilities of monocultures, and the unsustainability of plunder.

What, then, are we supposed to do? I have given a few public talks on my work exploring the connection between fake science and right-wing populism. I find that especially students in the audience then want to know what action they could take. There's not a lot that they can do as individuals. Recycling is well and good, but it does next to nothing to arrest environmental degradation. Resisting right-wing lies is useful, but I don't think supporting liberal alternatives is a lasting solution. Somehow we need a more deeply political response that goes beyond isolated individual actions and hoping for technological fixes.

And there, I confess, I run out of ideas. The culture of science, which I may have absorbed far too well, prompts us to back away from politics. If I speak as a scientist, I feel like I have to assume a position of impartiality, to present just the facts. In contrast, politics looks like a domain of lies, where misleading and vague promises are indispensable tools to hold together temporary coalitions. Within the professional class, we are constantly tempted by technocracy, by a desire to align truth with policy by avoiding the mess of politics and turning power over to the experts. So far, our technocratic tendencies have done more to corrupt expertise than to usher in any reign of truth.[54]

I could then lapse into my usual defeated cynicism. After all, I have done reasonably well for myself. I could shrug that too many people favor lies and work on creative ways to enslave one another, and that personally escaping that predicament is the best I can hope for.

Even that doesn't work. Much of what I have done in life is tied up with long-term projects that are much larger than myself, and that I would have hoped would continue long after I'm dead. I teach some students whom I hope will appreciate science better, some whom I help become scientists, and some whom I prompt to ask questions about weirdness and the nature of science. I publish papers about science and weirdness; few read them, but I still vaguely hope that they advance the cause of knowledge. What is the point of all that if civilization collapses?[55]

It seems I still cling to a vestigial hope of progress: the advancement of knowledge, if nothing else. When I was younger, I would have supported a more robust notion of human progress. I have abandoned most of that. I have no spectacular story of a loss of faith; I have gradually lost confidence until my youthful enthusiasm now seems embarrassing. Ironically, the classic secular liberal preoccupation with criticizing religious forms of weirdness had a lot to do with my eroding confidence. Confronting many religious arguments, I have been convinced by a few of them, particularly those opposing the universalist ambitions of secularism and liberal individualism.[56] Moreover, for the past two decades, the realization that our civilization won't properly respond to our climate crisis has been settling on me and driving me slightly mad. Even the modest hopes of progress that gave meaning to my long-term projects are slipping out of reach.

In Weird Science, students looking into claims of space aliens often discover the Fermi Paradox. *Star Trek* is never going to happen: humans are not at all suited for space, the distance between stars is too immense, and the resource costs for sending humans across the galaxy are absurdly large. But a technological civilization could still colonize the galaxy by sending artificially intelligent robots to neighboring stars with planets. These robots would travel at speeds very slow compared to the speed of light. They could then replicate themselves using resources mined at their destinations and spread slowly across the galaxy. If technological civilizations were abundant in our galaxy, it seems that their robotic representatives should already be here. Why not?

One answer the students find ties into our discussions on existential threats. Maybe technological civilizations are exceedingly rare, and one reason is that most destroy themselves by nuclear war, environmental catastrophe, or through any of the long list of existential threats we can imagine once their technology achieves a scale that has planetary consequences.[57] We don't know. The Fermi Paradox is very helpful in exposing the level of our ignorance about such questions.

Ignorance, perhaps, is not all that bad. I'm not an optimist by nature, and I can easily get into black moods about our future. But it is still possible that we can stumble upon a technological fix to slow down and limit the damage from climate change. Our luck might continue to hold with nuclear war. And in any time we buy, we may have an opportunity to cultivate less corrupt elites and a more competent professional class. We might stop confusing plunder with the limited forms of progress that are still available to us. We might even be able to crawl out of our current trap of individual instrumental rationality and collective insanity. We might deal with the wounds inflicted by civilization with something other than high-tech Band-Aids.

Conflicts between interests will never vanish. I have never been attracted to religious myths of paradise, and I am still doing my best to escape the secular myth of progress. If we have a future, given time enough, our world will change so much that my aspirations for the long term will become completely irrelevant. But that, for most of us, still has to be better than going out in a blaze of our own stupidity. And real science, rather than weirdness, will be a lot more helpful for realizing any decent future.

Grand Narratives

There once was a leaderless collective of gods who were not entirely satis-fied with their blissful existence contemplating their essential Oneness in the Realm of Eternity. Eternal bliss had its good points, no doubt. It was bliss, after all. But there was something vaguely bothersome about noticing the Realm of Change every now and then. After all, the inferior Realm seemed to be good for nothing but providing an illusion to escape from. Most of the gods in the leaderless collective thought they should be more productive, what with all their amazing powers and all that.

The multiverse was being its usual chaotic self. One more spacetime bubble had inflated as a random fluctuation from the pre-geometric quantum foam, and it looked like this was one of the more stable ones. It had already cooled down enough for galaxies to form; its disorder grew but could not keep pace with the maximum possible disorder, opening up space for spontaneous order.

Some of the gods in the leaderless collective thought that they might devise a game set in the Realm of Change, a diversion for those within Eternity. The game would be populated with agents who would do exciting things. These agents would be similar enough to the gods that the players and spectators would be able to identify with the agents. And the agents would have a large measure of independence, so that they could surprise and delight the watch-ing and playing gods.

As the spacetime bubble changed, the first generation of stars formed and died, their explosive deaths spurring the birth of successive generations of stars and enriching them with heavy elements. On some of the rockier planets and moons constituted from the leftovers of star formation, there were conditions that could sustain incredibly complex chemical processes.

One of the more artistically oriented gods in the leaderless collective had an idea: they should limit the life span of the game agents they would make in their own image. After all, the limitations of a medium were a spur to creativity. If the agents were immortal, they would be cursed with the ennui of Eternity. Every finite accomplishment would be worthless, every finite act meaningless, with an infinity of life always extending in front of them. All purposes would be fulfilled and every hope dashed. Immortality would mean that the taste of life would eventually become ashes in the mouths of the agents.

The bacteria emerged first, and remained the dominant form of life. A few strains, however, fell into a way of life that favored close cooperation. They eventually led to multicellular organisms. Evolution produced blind alleys and failed experiments, and life rode out planetary catastrophes. Dinosaurs stomped around.

A subset of the leaderless collective formed a Design Committee to develop ideas about the game and the agents in the game. After some study, a majority of the members of the Design Committee felt that if the agents were to be obliterated at the end of their spans, their lives would again become meaningless. All striving would be a sound and fury swallowed by death. So they proposed that agents should be made of two parts. There would be a body composed of the native matter of the Realm of Change, but there would also be an immortal bit that was a spark of Eternity. The immortal part would also be endowed with paranormal powers over the mere matter that was subject to Change.

After the dinosaurs had a stroke of remarkably bad luck, mammals radiated into the megafauna niches, though they never had quite the same tendency toward the gigantic as the dinosaurs they replaced. The bird branch of the dinosaurs was fortunate and survived. The bacteria kept thriving.

The artist who had thought up death was not happy: the two-part agent seemed to be a kludge, an affront to the integrity of the original concept. But the other gods persuaded her that there could be no perfection in the Realm of Change, and that, all things considered, the combination of a body and a spirit was a rather elegant solution.

And it came to pass that evolution coughed up a relatively hairless, big-brained ape that walked upright, made things with its hands, and could communicate well enough to incessantly gossip about its friends and enemies. The ape got lucky and spread through the continents, eventually starting to act more like a predator than the prey its ancestors had been. For many thousands of years the bands of humans roamed the plains, foraging, drinking, telling stories, and having a raucous good time.

The Design Committee reported back to the leaderless collective that the fate in Eternity of the immortal part of the game agents had to be decided, and that this was beyond their brief. The gods then decided to split into Working Groups to consider the question.

One of the greatest geniuses among the humans, and also one of the greatest fools, invented agriculture. Fortunately, her discovery was quickly forgotten. But others kept reinventing it. Eventually, most humans found that they were enslaved by agrarian empires. Some among the elite men in the empires and city states found that they had free time. Most indulged in court intrigue, but a few tried to understand the motions of the stars.

The Working Groups began to report back. One proposed that the immortal spirits could contemplate the mathematical Truths that resided in Eternity. The spirits could even figure out new games for the gods to play. Another Working Group suggested that the spirits of the dead should be given seventy-two virgins, though a minority report said that Eternal torture would be more amusing for the gods to watch. One Group thought it would make most sense if the spirits of dead ancestors were able to communicate with and guide their descendants. Another Group said that the dead should be considered lesser gods and should partake, as much as their capacity allowed, in the bliss of the truly Eternal.

It came to pass that the humans improved their technology and had a scientific revolution. States descended from empires harnessed fossil fuels, industrial-ized, and swallowed the world. Their scientists discovered that the stars were ongoing thermonuclear explosions. The humans cleared the forests, burned their fuels, and heated the skies. They made nuclear bombs. They figured out how to genetically engineer bioweapons that could destroy everyone. They made thinking machines and worried about whether the machines would wipe humans out.

After a prolonged conference and heated discussions, the leaderless collective organized itself, electing a Chairgod, a Vice Chairgod, and various Officers. The new Directorate decided to combine the most popular proposals, setting aside thirty-nine different Subrealms in Eternity and sorting the dead into them. They would invent something called Morality to keep score in the game and determine the fate of the dead. Each spirit would end up in an appropriate Subrealm according to whether they were good or evil, whether they had developed their paranormal powers, and whether they had contemplated enough mathematics. Different teams of gods, organized into Subdirectorates, would administer each Subrealm and make decisions about further partitioning their Subrealm in consultation with an overall Advisory Committee.

The humans blew themselves up. Climate upheavals exacerbated international tensions and created waves of climate refugees, overwhelming the human states. A nuclear exchange began with one of the peripheral vassal states but drew in the decaying central empire whose elites were worried about challenges from economic competitors. A remnant of humans survived, mostly attached to the bunkers billionaires had built. But with a broader biological collapse going on, they could not sustain themselves. Humans went extinct. Their planet was fine. Bacteria flourished.

Many of the gods in the original leaderless collective were unhappy about how the project had developed, and so they gave up and returned to their state of bliss. There were conflicts among the Subdirectorates; one breakaway faction tried to take control of the project. This faction had been closely involved in devising Morality; they became too invested in their own creation and came to think that there should be only two Subrealms rather than thir-ty-nine, just a heaven and a hell. Quarrels broke out, more of the gods lost interest, and there was growing administrative gridlock.

Stars kept dying and being born, but the fabric of this spacetime bubble became stretched and old. Eventually, new stars no longer formed. New intelligent life and technological civilizations still appeared, at the usual rate of about once every ten billion years in each galaxy. They almost always destroyed them-selves within a few centuries of figuring out how the stars burned. Dark energy dominated the ever-accelerating expansion of the spacetime bubble, until the galaxies became too far apart and too faint. In fact, in this bubble, the dark energy came in the form of phantom energy, so that after a few trillion years, a big rip occurred, and the spacetime bubble ceased to exist as all distances became infinite.

The game project in ruins despite the best efforts of the Chairgod, the re-maining gods drifted back toward the passionless Perfection of Eternal bliss. They contemplated Eternal things, and forgot that they had even entertained the insane possibility of sullying their Perfection by playing with lowly and finite matter.

THE END OF WEIRDNESS

I don't have any grand unified theory of weirdness to offer. There might have been such a theory, if science had some kind of essence that transcended mundane realities: a method, a logic, a reflection of a Platonic form of Reason. In that case, deviations from that essence would result in weirdness. But I don't think so. There is no scientific method into which we can feed evidence, work the machinery, and produce truth as a product. Human reason is a thoroughly fallible, evolved collection of tools that have helped us pursue our interests. Truth is not beauty, and both are a bit of a mess.

I can't point to any damning flaw shared by all forms of weirdness. Psychic powers, I think, are not real. But studying psychic claims produces no end of ambiguities and uncertainties, as well as obvious examples of fake science. There is nothing about parapsychology that is blatantly misconceived. The alleged feats of psychics and miracle workers violate the laws of physics, which is a good reason to be very suspicious. But we can still investigate psychic claims. Although I think that parapsychology has failed to make a convincing case, it is not the only experimental science that has lately found itself in a muddle.

Weird claims have roots in our common human psychology. They then get encrusted in our traditions, including not just our religions but also our most sophisticated intellectual enterprises. This includes the academic mainstream. Even our natural sciences still preserve echoes of occult ways of knowing. Skeptics and scientists can come close to promoting our own brand of weirdness, especially when we uphold a heroic image of science, treat reason as a transcendent ideal, and embrace myths of progress.

People have many purposes in life, and a devotion to truth above every other interest would be a form of fanaticism. Similarly, our institutions serve many purposes, and achieving accurate descriptions of the world is rarely their first priority. Our worldly forms of reason, closely tethered to the reproductive interests that shape us, do not demand that we expunge ourselves of false beliefs. When skeptics and scientists hope for a utopia where reason overcomes superstition, we are liable to imagine technocracies, reigns of rationally determined moral principles, and science-fiction dreamscapes. Technocracy is not to my taste. Objective moral facts fixed by reason are as much items of weirdness as the gods and demons of traditional religions. The dreamscapes, though, are fine. I have no cause to complain about imagination. Science fiction is very much my taste.

Scientific institutions are entangled with politics. That need not be a problem. Our post-truth chaos and the existential threats on our horizon suggest that any viable technological civilization needs to make much better use of scientific expertise. There is, however, a real danger of support for science shading into an uncritical endorsement of the interests of professionals. In such a political context, a heroic image of science and a transcendent conception of reason can threaten to turn science into an equivalent of priestly authority. Perhaps I am overly cautious. But when I think about the institutional pathologies exhibited by varieties of creationism, I notice little that is not also found, to some degree, in mainstream academic life. Without a method that guarantees truth, only a careful weighing of all the relevant arguments, as best as we can manage, keeps mainstream science from behaving like its fake counterparts.

What, then, can be done about weirdness? Many scientists and skeptics think of fake science and the paranormal as a public nuisance. We'd like to put an end to weirdness, though we suspect that isn't practically possible. I suspect the situation is more complicated. The boundaries of science are always fuzzy, and the arguments are never as conclusive as I might like. In principle, what seems to be weirdness and what is textbook science can always shift with new findings. And yet, we shouldn't expect many radical changes. Even with our complex social ecologies full of ever-conflicting interests, we might reach enough of a consensus about the harms of ignoring science. Uncertainty cannot mean paralysis; we cannot let climate change denial, creationism, or alternative medicine run rampant.

I don't have a five-point plan to combat weirdness and elevate real science. Those of us who teach science know that a lot of what we do is an art. Allegedly scientific research on education is notorious for its oversimplifications and susceptibility to changing fashions. Fighting fake science is a political matter, and negotiating political interests is even less likely to leave us with simple recipes. So I expect that those of us who are fans of science will keep trying to resist popular forms of weirdness in a variety of ways. Scientists will speak out, as a contribution to public education. Skeptics will publish magazines, hold forth online, and try to get our viewpoint into the mass media. Maybe, just maybe, scientific institutions will come to recognize that a public dialogue about science and reality cannot be a one-way transmission of scientific truth from scientists to a passive audience.

What I might recommend, however, is that scientists and skeptics can make use of debates about weirdness as a way to appeal to the intellectual curiosity of many people who are disengaged from science. My experience teaching Weird Science to students from all backgrounds suggests that there are many who otherwise couldn't care less about the philosophy of science who can be reached through their curiosity about the paranormal. They get caught up in exploring questions about weirdness, about reality, and how we figure out whom to trust. There is no guarantee that curiosity about weird claims will result in a more skeptical, mainstream scientific point of view. The only time I have been nominated for Educator of the Year at my university was when a student with creationist sympathies found that some philosophical arguments that came up in Weird Science could be deployed to fashion an apologetic for intelligent design, and wrote his letter of nomination in a burst of enthusiasm. By and large, however, even without my prompting, I find that my students become more skeptical as they learn more. The creationist student who nominated me let me know, some years after he graduated, that he had continued exploring, and that he could no longer sustain a fundamentalist faith or indeed any conventional religion.

None of this heralds the end of weirdness. But if mainstream scientific descriptions of the world are mostly accurate, we should also expect that disciplined reflection on weirdness will often lead to more skepticism. In any case, such reflection is fun. I have made it a major part of my career, and I have been fortunate to offer opportunities for reflection to my students. The Weird Science course itself will come to an end, perhaps sooner than I would have hoped, since my university is now suffering from the effects of a pandemic on top of a permanent financial crisis. Protecting what we can of a shrinking physics program may mean sacrificing our more liberal-artsy course offerings. But there are other universities, where other faculty explore questions about science and fake science. In the long run, that might not help either, because most universities are dying. Still, old institutions falter, but if we're lucky and if civilization lasts, we will build new institutions. They might, by design or by accident, provide opportunities for reflection. And thinking about weirdness will still be fun.

GRATITUDE

I owe much to my students in Weird Science over the last two decades. I eventually forget the details, so that all that remains might be memories such as a time when a psychology major made a telling observation. Nonetheless, there are echoes of much that happened in the classroom in this book.

I instinctually approach anyone who thinks they're management, such as university administrators, with suspicion. But for whatever reason, no chairs or deans or provosts or anyone have suggested that I should drop my obsession with weirdness and behave like a proper physicist. Perhaps this is because we're a small public liberal arts university in the middle of nowhere, and there is little prospect that we can join the perpetual hunt for grant money. In any case, I am grateful for the opportunity I have found to pursue my unconventional interests, and I should acknowledge that this depends on the efforts of many people who work to keep the lights on.

Amy Bix, Stefaan Blancke, Maarten Boudry, and Tom Marshall read and offered criticism on drafts of chapters. I am thankful for their help. Writing a book is peculiar. It's mostly solitary work, with feedback from a handful of people, but I also feel as if I am communicating with thousands of others separated in space and time.

NOTES

Chapter 1

1. Spiritualists often thought of themselves as investigating spiritual realities in a scientific manner. Janet Oppenheim, *The Other World: Spiritualism and Psychical Research in England, 1850–1914* (Cambridge: Cambridge University Press, 1985).

2. Kendrick Frazier, "History of CSICOP," in Gordon Stein, ed., *The Encyclopedia of the Paranormal* (Amherst: Prometheus Books, 1996).

3. Among today's skeptics, the diagnostic approach has evolved to include lists of fallacies and cognitive biases. For example, Steve Novella, *The Skeptics' Guide to the Universe: How to Know What's Really Real in a World Increasingly Full of Fake* (New York: Grand Central Publishing, 2018); Charles M. Wynn and Arthur W. Wiggins, *Quantum Leaps in the Wrong Direction: Where Real Science Ends—And Pseudoscience Begins* (2nd edition, New York: Oxford University Press, 2017).

4. For criticism of criteria such as "methodological naturalism" in this context, see Maarten Boudry, Stefaan Blancke, and Johan Braeckman, "Grist to the Mill of Anti-evolutionism: The Failed Strategy of Ruling the Supernatural Out of Science by Philosophical Fiat," *Science & Education* 21: 1151–1165 (2012), doi:10.1007/s11191-012-9446-8; Jeffrey Koperski, "Two Bad Ways to Attack Intelligent Design and Two Good Ones," *Zygon* 43(2): 433–439 (2008), doi:10.1111/j.1467-9744.2008.00926.x; Sandy C. Boucher, "Methodological Naturalism in the Sciences," *International Journal for Philosophy of Religion* 88: 57–80 (2020), doi:10.1007/s11153-019-09728-9.

5. Andrew A. Snelling, "Radioisotope Dating of Grand Canyon Rocks: Another Devastating Failure for Long-Age Geology," *Acts & Facts* 33(10) (2004).

6. Karl Popper, *Conjectures and Refutations: The Growth of Scientific Knowledge* (New York: Routledge and Kegan Paul, 1963).

7. Michael J. Benton, *Dinosaurs Rediscovered: The Scientific Revolution in Paleontology* (London: Thames & Hudson, 2019).

8. Carol E. Colaninno, "Anxiety: Should Children Be Afraid of Bigfoot?" *Skeptical Inquirer* 44(2): 51–52 (2020).

9. Robert J. Richards and Lorraine Daston, eds., *Kuhn's* Structure of Scientific Revolutions *at Fifty: Reflections on a Science Classic* (Chicago: University of Chicago Press, 2016). George A. Reisch, *The Politics of Paradigms: Thomas S. Kuhn, James B. Conant, and the Cold War "Struggle for Men's Minds"* (Albany: State University of New York Press, 2019).

10. Keith M. Ashman and Philip S. Baringer, eds., *After the Science Wars* (London: Routledge, 2001). Keith M. Parsons, *The Science Wars: Debating Scientific Knowledge and Technology* (Amherst: Prometheus Books, 2003).

11. Even some authors who construct checklists to diagnose fake science are aware of the limitations of falsificationism. For example, I used to assign a textbook for Weird Science, Theodore Schick, Jr. and Lewis Vaughn, *How to Think About Weird Things: Critical Thinking for a New Age* (8th edition, New York: McGraw-Hill Education, 2019), which is lukewarm about falsificationism.

12. Daniel Loxton and Donald R. Prothero, *Abominable Science!: Origins of the Yeti, Nessie, and Other Famous Cryptids* (New York: Columbia University Press, 2013).

13. Mike Dupler, *On the Trail of Bigfoot: Tracking the Enigmatic Giants of the Forest* (Newburyport: New Page Books, 2020), chapter 6.

14. Evasive strategies, however, are common in fake science, and the distinction between an immunizing maneuver and a natural extension of a claim can be hard to draw. Maarten Boudry and Johan Braeckman, "Immunizing strategies and epistemic defense mechanisms," *Philosophia* 39: 145–161 (2011), doi:10.1007/s11406-010-9254-9.

15. Timothy H. Heaton, "Recent Developments in Young-Earth Creationist Geology," *Science & Education* 18(10): 1341–1358 (2009), doi:10.1007/s11191-008-9162-6.

16. Our current best cosmological models also require dark matter, and some other data such as gravitational lensing also are most consistent with dark matter. Rotation speed data, however, is the best direct evidence. Barbara

Ryden, *Introduction to Cosmology* (2nd edition, Cambridge: Cambridge University Press, 2017), 123–140.

17. The OPERA collaboration. Chad Orzel, "Scientific Failure as a Public Good: Illustrating the Process of Science and Its Contrast with Pseudoscience," in Allison B. Kaufman and James C. Kaufman, eds., *Pseudoscience: The Conspiracy Against Science* (Cambridge: The MIT Press, 2018).

18. Modified gravity ideas, though minority views, are still in play. Kyu-Hyun Chae et al., "Testing the Strong Equivalence Principle: Detection of the External Field Effect in Rotationally Supported Galaxies," *The Astrophysical Journal* 904(1): 51–70 (2020), doi:10.3847/1538-4357/abbb96. Moritz Platscher, Juri Smirnov, Sven Meyer, and Matthias Bartelmann, "Long Range Effects in Gravity Theories with Vainshtein Screening," *Journal of Cosmology and Astroparticle Physics,* 12: 009 (2018), doi:10.1088/1475-7516/2018/12/009. For a philosophical perspective, see David Merritt, *A Philosophical Approach to MOND* (Cambridge: Cambridge University Press, 2020).

19. Popper discussed statements such as "All swans are white"; Karl Popper, *The Logic of Scientific Discovery* (London: Routledge Classics, 2002). But I like crows a lot more than swans.

20. For philosophers' views on "scientific method" in science textbooks, see Brian A. Woodcock, "'The Scientific Method' as Myth and Ideal," *Science & Education* 23: 2069–2093 (2014), doi:10.1007/s11191-014-9704-z; James Blachowicz, "How Science Textbooks Treat Scientific Method: A Philosopher's Perspective," *The British Journal for the Philosophy of Science,* 60(2): 303–344 (2009), doi:10.1093/bjps/axp011.

21. Jan Sprenger and Stefan Hartmann, *Bayesian Philosophy of Science: Variations on a Theme by the Reverend Thomas Bayes* (Oxford: Oxford University Press, 2019).

22. Richard Dietz, ed., *Vagueness and Rationality in Language Use and Cognition* (Cham: Springer Nature, 2019). Contributors point out the value of vagueness, but also argue that some long-standing criticisms of Bayesian reasoning for false precision can be bypassed. I am not as confident, especially where initial probabilities are concerned.

23. Let E stand for the evidence, that an observed crow was black, and let H be a hypothesis about the distribution of colors. Our model has $P(E) = 1/2$, and the starting probabilities are $P(H) = (1/2)^N$, where there are N crows in the population. Each hypothesis has an individual crow being either black or nonblack, so that $P(E|H) = 1$ or 0. If one crow is sampled and $P(E|H)$ was 0, that hypothesis will be eliminated. If $P(E|H) = 1$ instead, by Bayes's theorem $P(H|E) = P(H)P(E|H)/P(E) = (1/2)^{N-1}$. This is the same as the starting

position but now with a population of $N-1$. The evidence, therefore, says nothing about the unobserved crow population.

24. P. W. Bridgman, "New Vistas for Intelligence," in Karl K. Darrow, ed., *Physical Science and Human Values: A Symposium* (Princeton: Princeton University Press, 1947), 144.

25. Richard Horton, "Offline: What Is Medicine's 5 sigma?" *The Lancet* 385: 1380 (2015), doi:10.1016/S0140-6736(15)60696-1. Stuart Ritchie, *Science Fictions: How Fraud, Bias, Negligence, and Hype Undermine the Search for Truth* (New York: Metropolitan Books, 2020). Brian M. Hughes, *Psychology in Crisis* (London: Palgrave MacMillan, 2018). I don't, however, want to claim that replication is a necessary feature of science. See Allan Franklin and Ronald Laymon, *Once Can Be Enough: Decisive Experiments, No Replication Required* (Cham: Springer Nature, 2020). Neither do I want to suggest that tighter statistical significance requirements is a technical fix that will solve all problems. Valentin Amrhein, Sander Greenland, and Blake McShane, "Scientists Rise Up Against Statistical Significance," *Nature* 567: 305–307 (2019), doi:10.1038/d41586-019-00857-9.

26. Naturally, I started out as a fan of the logical positivists; one of my early favorites was Alfred Jules Ayer, *Language, Truth and Logic* (New York: Dover Publications, 1952).

27. R.D. Rosenkrantz, ed., *E.T. Jaynes: Papers on Probability, Statistics and Statistical Physics* (Dordrecht: Kluwer Academic Publishers, 1989).

28. For some curious paradoxes, see William Eckhardt, *Paradoxes in Probability Theory* (Dordrecht: Springer, 2013). In objective Bayesian inference, the main and probably unsolvable problem is determining initial probabilities for continuous distributions.

29. Werner Gitt, *Did God Use Evolution? Observations from a Scientist of Faith* (Green Forest: Master Books, 2006). Dave A. Schoch, *The Assumptions Behind the Theory of Evolution: Why We Are Taught Assumptions as Codified Facts of Science* (Bloomington: AuthorHouse, 2010).

30. General approaches to induction run into issues with noncomputability, though approximations are possible; Ming Li and Paul M.B. Vitányi, "Inductive Reasoning and Kolmogorov Complexity," *Journal of Computer and System Sciences* 44: 343–384 (1992), doi:10.1016/0022-0000(92)90026-F. With combinations of randomness and algorithms, cases of guaranteed failure need not happen; Taner Edis, "How Gödel's Theorem Supports the Possibility of Machine Intelligence," *Minds and Machines* 8: 251–263 (1998), doi:10.1023/A:1008233720449. Neural networks have similar issues: Herbert Wiklicky, "On the Non-Existence of a Universal Learning Algorithm for Recurrent Neural Networks," *Advances in Neural Information Processing*

Systems 6: 431–436 (1994).

31. Susan Haack, *Evidence and Inquiry: A Pragmatist Reconstruction of Epistemology* (expanded edition, Amherst: Prometheus Books, 2009).

32. My description owes something to the "web of belief" metaphor used by W.V. Quine: W.V. Quine and J.S. Ullian, *The Web of Belief* (2nd edition, New York: McGraw-Hill, 1978), though I don't agree with the radical holism often associated with such a view. It's too bad skeptics and scientists ended up approaching fake science with a rhetoric derived more from Popper rather than Quine or Haack.

33. Due to public interest in such matters, it's possible to find less technical explanations. For example, Sabine Hossenfelder and Stacy S. McGaugh, "Is Dark Matter Real?" *Scientific American* 319(2): 36–43 (2018); Ethan Siegel, "Why Supersymmetry May Be The Greatest Failed Prediction In Particle Physics History," *Forbes*, February 12, 2019, https://www.forbes.com/sites/startswithabang/2019/02/12/why-supersymmetry-may-be-the-greatest-failed-prediction-in-particle-physics-history/ (accessed December 21, 2020).

34. No less: we cannot erase the details and represent the reasons physicists have as a set of probabilities attached to hypotheses, the $P(H)$ in Bayesian statistics.

35. This is closely related to the view that human reason is fully natural, bounded, and evolved. For example, Konrad Talmont-Kaminski, *Religion as Magical Ideology: How the Supernatural Reflects Rationality* (Durham: Acumen, 2013), chapter 2. Such positions invite accusations of scientism; but see Johan Hietanen et al., "How *Not* to Criticise Scientism," *Metaphilosophy* 51: 522–547 (2020), doi:10.1111/meta.12443.

36. Some of the arguments in this section derive from Taner Edis, "Two Cheers for Scientism," in Maarten Boudry and Massimo Pigliucci, eds., *Science Unlimited? The Challenges of Scientism* (Chicago: University of Chicago Press, 2018).

37. Annie Besant and Charles W. Leadbeater, *Occult Chemistry: Clairvoyant Observations on the Chemical Elements* (London: Theosophical Publishing House, 1919).

38. Misrepresentations of quantum mechanics are very common in the literature defending psychic powers. See Jack A. Mroczkowski and Alexis P. Malozemoff, "Quantum Misuse in Psychic Literature," *Journal of Near-Death Studies*, 37: 131–154 (2019), doi:10.17514/JNDS-2019-37-3-p131-154, and responses to the article that make up the rest of the issue. But *if* psychic powers were real, perhaps a modified version of quantum mechanics could help describe psychic phenomena.

39. This is, naturally, a physicists' view. There are recurrent arguments that by the looser standards that prevail in psychology, there is at least reason to further investigate psychic claims. Etzel Cardeña, "The Experimental Evidence for Parapsychological Phenomena: A Review," *American Psychologist* 73(5): 663–677 (2018), doi:10.1037/amp0000236. For a skeptical response, see Arthur S. Reber and James E. Alcock, "Searching for the Impossible: Parapsychology's Elusive Quest" *American Psychologist* 75(3): 391–399 (2020), doi:10.1037/amp0000486.

40. Hasok Chang, "Presentist History for Pluralist Science," *Journal for General Philosophy of Science* (2020), doi:10.1007/s10838-020-09512-8. Hasok Chang, "Who Cares About the History of Science?" *The Royal Society Journal of the History of Science* 71(1): 91–107 (2017), doi:10.1098/rsnr.2016.0042. Laurent Loison, "Forms of Presentism in the History of Science. Rethinking the Project of Historical Epistemology," *Studies in History and Philosophy of Science* 60: 29–37 (2015), doi:10.1016/j.shpsa.2016.09.002.

41. Charles T. Tart, *The End of Materialism: How Evidence of the Paranormal Is Bringing Science and Spirit Together* (Oakland: New Harbinger Publications, 2009). Steve Taylor, "Moving Beyond Materialism: Can Transpersonal Psychology Contribute to Cultural Transformation?" *International Journal of Transpersonal Studies* 36(2): 147–159 (2017), doi:10.24972/ijts.2017.36.2.75. William A. Dembski and Jonathan Witt, *Intelligent Design Uncensored: An Easy-to-Understand Guide to the Controversy* (Downers Grove: InterVarsity Press, 2010).

42. Stefaan Blancke, Maarten Boudry, and Massimo Pigliucci, "Why do Irrational Beliefs Mimic Science? The Cultural Evolution of Pseudoscience," *Theoria* 83(1): 78–97 (2017), doi:10.1111/theo.12109. Maarten Boudry, "Diagnosing Pseudoscience—by Getting Rid of The Demarcation Problem," forthcoming. Fake science is probably inevitable in highly technological societies, where science becomes a status to aspire to. Michael D. Gordin, *The Pseudoscience Wars: Immanuel Velikovsky and the Birth of the Modern Fringe* (Chicago: University of Chicago Press, 2012).

43. See contributions in Massimo Pigliucci and Maarten Boudry, eds., *Philosophy of Pseudoscience: Reconsidering the Demarcation Problem* (Chicago: The University of Chicago Press, 2013); also, Stephen Law, "How Can We Tell Science from Pseudoscience?" in Kevin McCain and Kostas Kampourakis, eds., *What Is Scientific Knowledge?: An Introduction to Contemporary Epistemology of Science* (New York: Routledge, 2019). There are still attempts to find a compact demarcation criterion, such as Angelo Fasce, "What Do We Mean When We Speak of Pseudoscience? The Development of a Demarcation Criterion Based on the Analysis of Twenty–One Previous Attempts," *Disputatio* 7: 459–488 (2017), doi:10.5281/zenodo.1433737. However, I

think that the proposed criteria, to the extent that they are plausible, do not have much substance beyond highlighting the rejection of fake scientific claims by mainstream science.

44. Elliott Sober, "Let's Razor Ockham's Razor," in *From a Biological Point of View: Essays in Evolutionary Philosophy* (Cambridge: Cambridge University Press, 1994). Justifying Occam's razor regularly presents technical problems; see Tom F. Sterkenburg, "Solomonoff Prediction and Occam's Razor," *Philosophy of Science* 83: 459–479 (2016), doi:10.1086/687257. Still, with proper use of background knowledge in statistical models, a Bayesian version of simplicity can be useful. Thomas Blanchard, Tania Lombrozo, and Shaun Nichols, "Bayesian Occam's Razor Is a Razor of the People," *Cognitive Science* 42: 1345–1359 (2018), doi:10.1111/cogs.12573.

45. Some of the arguments in this section derive from Taner Edis, "From Creationism to Economics: How Far Should Analyses of Pseudoscience Extend?" *Mètode Science Studies Journal* 8: 141–147 (2018), doi:10.7203/metode.8.10001.

46. Stuart Vyse, *Believing in Magic: The Psychology of Superstition* (updated edition, New York: Oxford University Press, 2014).

47. My descriptions of Turkish creationism are derived from Taner Edis, "The Politics of Islamic Opposition to Evolution in Turkey," in C. Mackenzie Brown, ed., *Asian Religious Responses to Darwinism: Evolutionary Theories in Middle Eastern, South Asian, and East Asian Cultural Contexts* (Cham: Springer Nature, 2020); Taner Edis, "The Turkish Model of Islamic Creationism," *Almagest* 12: 40–65 (2021), doi:10.1484/J.ALMAGEST.5.125385.

48. Douglas Axe, *Undeniable: How Biology Confirms our Intuition That Life Is Designed* (New York: HarperCollins, 2016).

49. Lee McIntyre, *The Scientific Attitude: Defending Science from Denial, Fraud, and Pseudoscience* (Cambridge: The MIT Press, 2019).

50. There are few polls on creationism and applied scientists, though they show appreciable support: "Nearly Two-Thirds of Doctors Skeptical of Darwin's Theory of Evolution," *Discovery Institute*, May 31, 2005, https://www.discovery.org/a/2611/ (accessed December 21, 2020). More generally, however, applied scientists are politically and religiously much more conservative than natural scientists. Diego Gambetta and Steffen Hertog, *Engineers of Jihad: The Curious Connection between Violent Extremism and Education* (Princeton: Princeton University Press, 2016).

51. Jaime A. Teixeira da Silva and Judit Dobránszki, "Problems with Traditional Science Publishing and Finding a Wider Niche for Post-Publication Peer Review," *Accountability in Research* 22(1): 22–40 (2015), doi:10.1080/08989621.2014.899909. Richard Walker and Pascal Rocha da Silva, "Emerg-

ing Trends in Peer Review—A Survey," *Frontiers in Neuroscience* 9 (2015), doi:10.3389/fnins.2015.00169.

52. Hans Radder, ed., *The Commodification of Academic Research: Science and the Modern University,* (Pittsburgh: University of Pittsburgh Press, 2010).

53. My first instinct, when I first got involved with skepticism about paranormal claims, was to resist science studies. Today, while I may not agree with every detail, I think the overall vision of the social nature of science in, for example, Helen Longino, *The Fate of Knowledge* (Princeton: Princeton University Press, 2002), is largely correct. Having said that, social constructionism within science studies has been associated with science-denying mischief: Sven Ove Hansson, "Social Constructionism and Climate Science Denial," *European Journal for Philosophy of Science* 10: 37 (2020), doi:10.1007/s13194-020-00305-w. Striking a balance seems best: Susan Haack, *Defending Science—Within Reason: Between Scientism and Cynicism* (Amherst: Prometheus Books, 2007). A good introduction to science from a science studies point of view is Naomi Oreskes, *Why Trust Science?* (Princeton: Princeton University Press, 2019).

Chapter 2

1. Jason Haxton, *The Dibbuk Box* (Kirksville: Truman State University Press, 2011).

2. This stereotype should have been buried together with the Romantic era of the nineteenth century, but I still run into its echoes. For example, Curtis White, *The Science Delusion: Asking the Big Questions in a Culture of Easy Answers* (Brooklyn: Melville House 2013).

3. This is a long-standing objection to psychic claims, and still a centerpiece of skeptical criticism. Arthur S. Reber and James E. Alcock, "Searching for the Impossible: Parapsychology's Elusive Quest" *American Psychologist* 75(3): 391–399 (2020), doi:10.1037/amp0000486.

4. Sebastien Point, "Free Energy: When the Web Is Freewheeling," *Skeptical Inquirer* 42(1): 52–55 (2018).

5. A common creationist response is to accept that such events violate the laws of physics, but declare that the events were supernatural, and therefore everything is fine. For example, Shaun Doyle, "Too Much Heat in Noah's Flood?" *Creation.com* July 25, 2020, https://creation.com/flood-heat-problem (accessed December 24, 2020).

6. Or speculate about exotic physics or engage in quantum obfuscation. For example, Elizabeth A. Rauscher and Russell Targ, "The Speed of Thought: Investigation of a Complex Space-Time Metric to Describe Psychic Phe-

NOTES

nomena," *Journal of Scientific Exploration,* 15(3): 331–354 (2001); Bryan J. Williams, "Reassessing the 'Impossible': A Critical Commentary on Reber and Alcock's 'Why Parapsychological Claims Cannot Be True'," *Journal of Scientific Exploration* 33(4): 599–616 (2019), doi:10.31275/2019/1667.

7. Crank material on free energy through zero-point energy is easy to find; a useful starting point is http://www.zpenergy.com (accessed February 27, 2021).

8. For example, Paul R. Hill, *Unconventional Flying Objects: A Former NASA Scientist Explains How UFOs Really Work* (revised edition, Charlottesville: Hampton Roads Publishing, 2014). The "scientist" is an engineer; physicists better appreciate how unlikely antigravity is. Naturally, there is no shortage of cranks working on antigravity. David Hatcher Childress, ed., *The Anti-Gravity Files* (Kempton: Adventures Unlimited Press, 2017).

9. All this is standard physics, and the subject of many popular physics books. To my introductory physics class for nonscience majors, I assign Neil De-Grasse Tyson, *Astrophysics for People in a Hurry* (New York: W.W. Norton, 2017).

10. Stephen M. Stigler, "Stigler's Law of Eponymy," in Thomas F. Gieryn, ed., *Science and Social Structure: A Festschrift for Robert K. Merton* (New York: New York Academy of Sciences, 1980).

11. Dwight E. Neuenschwander, *Emmy Noether's Wonderful Theorem* (revised edition, Baltimore: Johns Hopkins University Press, 2017). In quantum mechanics, individual results of identical experiments are random and hence may be different, but their statistical distribution will be the same. See also Marc Lange, *Because Without Cause: Non-Causal Explanations in Science and Mathematics* (New York: Oxford University Press, 2017), 46–130.

12. Walter Greiner and Berndt Müller, *Gauge Theory of Weak Interactions* (fourth edition, Berlin: Springer-Verlag, 2009), 11–12.

13. Barbara Ryden, *Introduction to Cosmology* (2nd edition, Cambridge: Cambridge University Press, 2017). Standard cosmology assumes homogeneity and isotropy at large distance scales, but there are observations, such as "dark flow" that are inconsistent with these symmetries. F. Atrio-Barandela, A. Kashlinsky, H. Ebeling, D. J. Fixsen, and D. Kocevski "Probing the Dark Flow Signal in *WMAP* 9 -Year And *Planck* Cosmic Microwave Background Maps," *The Astrophysical Journal* 810(2): 143–159 (2015), doi:10.1088/0004-637X/810/2/143.

14. J.D. Vergados, *The Standard Model and Beyond* (Singapore: World Scientific, 2017), 113–134. At an undergraduate level, Masud Chaichian, Hugo Perez Rojas, and Anca Tureanu, *Basic Concepts in Physics: From the Cosmos to Quarks* (Berlin: Springer, 2013) is a good source.

15. Losing an anchor in experiment can lead to serious problems with the cutting edge of theoretical physics. Sabine Hossenfelder, *Lost in Math: How Beauty Leads Physics Astray* (New York: Basic Books, 2018). David Lindley, *The Dream Universe: How Fundamental Physics Lost Its Way* (New York: Doubleday, 2020).

16. All this is, again, standard physics. For a nonmathematical introduction to quantum mechanics, see Alistair I. M. Rae, *Quantum Physics: A Beginner's Guide* (New York: Simon and Schuster, 2005).

17. Maarten Boudry, Michael Vlerick, and Taner Edis, "The End of Science? On Human Cognitive Limitations and How to Overcome Them," *Biology & Philosophy* 35:18 (2020), doi:10.1007/s10539-020-9734-7.

18. The famous $\Delta p \Delta x \geq \hbar/2$ is a combination of $p = \hbar k$, which is the relationship between momentum and wave number that is basic to quantum mechanics, and $\Delta k \Delta x \geq \frac{1}{2}$, which comes from waves as probability amplitudes. Beyond basic quantum mechanics, the stochastic approach to quantum mechanics produces a natural, nonmysterious explanation: Jussi Lindgren and Jukka Liukkonen, "The Heisenberg Uncertainty Principle as an Endogenous Equilibrium Property of Stochastic Optimal Control Systems in Quantum Mechanics," *Symmetry* 12: 1533 (2020), doi:10.3390/sym12091533.

19. There are many philosophical interpretations of quantum mechanics, which have no difference whatsoever in terms of experimental predictions. Physicists often set all such interpretations aside and adopt a "shut up and calculate" approach. Not everyone is happy about that, and it still might be possible to engage in interpretation without mystification; Steven French and Juha Saatsi, eds., *Scientific Realism and the Quantum* (Oxford: Oxford University Press, 2020).

20. The measurement problem is often posed in terms of the role of the observer in quantum mechanics, but the way I describe it highlights the close connection between the measurement problem in quantum mechanics and the question of macroscopic irreversibility in statistical physics. Some versions of the measurement problem are due to conceptual confusion about state reduction and randomness; Berthold-Georg Englert, "On Quantum Theory," *The European Physical Journal D* 67: 238 (2013), doi:10.1140/epjd/e2013-40486-5. Alternatively, slight modifications to standard quantum mechanics such as the GRW collapse theory also lead to formulations without observers. Detlef Dürr and Dustin Lazarovici, *Understanding Quantum Mechanics: The World According to Modern Quantum Foundations* (Cham: Springer Nature, 2020).

21. Jack A. Mroczkowski and Alexis P. Malozemoff, "Quantum Misuse in Psychic Literature," *Journal of Near-Death Studies*, 37: 131–154 (2019),

doi:10.17514/JNDS-2019-37-3-p131-154. There will likely always be efforts to salvage something from quantum mechanics to apply to paranormal concepts of consciousness; for example, Bernardo Kastrup, "Reasonable Inferences From Quantum Mechanics: A Response to 'Quantum Misuse in Psychic Literature," *Journal of Near-Death Studies*, 37: 185–200 (2019), doi:10.17514/JNDS-2019-37-3-p185-200. Far from reflecting the current state of play in work on the foundations of quantum mechanics, as Kastrup claims, these are fringe ideas.

22. The classic in this genre is Deepak Chopra, *Quantum Healing: Exploring the Frontiers of Mind/Body Medicine* (revised edition, New York: Bantam Books, 2015).

23. The American one, https://quantumuniversity.com, is an online university specializing in "Holistic, Natural, and Integrative Medicine." I counted one and a half genuine physicists among their faculty. There is also a Quantum University in India, http://www.quantumuniversity.edu.in, that emphasizes IT.

24. Jennifer Burwell, *Quantum Language and the Migration of Scientific Concepts* (Cambridge: The MIT Press, 2018). Sadri Hassani, "Sources Of Quantum Voodooism," *Skeptical Inquirer* 44(6): 43–47 (2020).

25. The most prominent example I have come across is Amit Goswami, a retired physicist who produces no end of New Agey books with "quantum" in the title, recently including *Quantum Politics* and the *Quantum Science of Happiness*. I can't keep up with all of Goswami's output, but I can vouch for *The Physics of the Soul: The Quantum Book of Living, Dying, Reincarnation, and Immortality* (Charlottesville: Hampton Roads Publishing, 2001) as a classic of weirdness. No, "quantum nonlocality" does not mean a memory residing out of space and time that is the basis for a Hindu concept of immortality. Goswami should know better, since he's also the author of a fairly standard textbook, *Quantum Mechanics* (New York: McGraw-Hill Education, 1991). And yet, that's what he argues.

26. I've contributed, as a token skeptic, to two volumes which include monotheistic scientists, primarily with a Jewish background, arguing for some version of creationism. Joseph Seckbach and Richard Gordon, eds., *Theology and Science: From Genesis to Astrobiology* (Singapore: World Scientific, 2019); Joseph Seckbach and Richard Gordon, eds., *Divine Action and Natural Selection: Science, Faith and Evolution* (Singapore: World Scientific, 2009).

27. Meera Nanda, *Prophets Facing Backward: Postmodern Critiques of Science and Hindu Nationalism in India* (Piscataway: Rutgers University Press, 2003). Robert M. Geraci, *Temples of Modernity: Nationalism, Hinduism,*

and Transhumanism in South Indian Science (Lanham: Lexington Books, 2018).

28. A rule of thumb in these cases is that the typical size of statistical fluctuations is about the square root of the population. So if we have a macroscopic sample of about 10^{24} radioactive nuclei, the relative size of the fluctuations, which is the ratio of the fluctuations to the expected value, is about $10^{12}/10^{24}$ = 10^{-12}, one part in a trillion. The deterministic equation that gives a smooth exponential decay characterized by a half-life is completely reliable.

29. I am describing the randomness in quantum mechanics here by drawing a contrast between the kind of unpredictability associated with dynamical chaos, with its exponential divergence from initial conditions, and true randomness. Note that there are still some open questions concerning quantum randomness: see Andrei Khrennikov, *Probability and Randomness: Quantum versus Classical* (Singapore: World Scientific, 2016), 205–209. It is also not possible to strictly guarantee that quantum mechanics produces true randomness; Karl Svozil, *Physical (A)Causality: Determinism, Randomness and Uncaused Events* (Cham: Springer Nature, 2018). At present, however, all our evidence indicates so.

30. Robert John Russell, "Divine Action and Quantum Mechanics: A Fresh Assessment," in Fount LeRon Shults, Nancey C. Murphy and Robert John Russell, eds., *Philosophy, Science and Divine Action* (Boston: Brill, 2009). For technical criticisms, see Timothy Sansbury, "The False Promise of Quantum Mechanics," *Zygon* 42(1): 111–121 (2007), doi:10.1111/j.1467-9744.2006.00808.x; Jeffrey Koperski, "Divine Action and the Quantum Amplification Problem," *Theology and Science* 13(4), 379–394 (2015), doi:1 0.1080/14746700.2015.1082872.

31. Taner Edis, "Cosmic Conspiracy Theories: How Theologies Evade Science," in Joseph Seckbach and Richard Gordon, eds., *Theology and Science: From Genesis to Astrobiology* (Singapore: World Scientific, 2019), 154–157.

32. Charles T. Tart, *The End of Materialism: How Evidence of the Paranormal Is Bringing Science and Spirit Together* (Oakland: New Harbinger Publications, 2009). William A. Dembski and Jonathan Witt, *Intelligent Design Uncensored: An Easy-to-Understand Guide to the Controversy* (Downers Grove: InterVarsity Press, 2010).

33. Adam Auch, "Hype, Argumentation, and Scientific Dissemination," in Peter Vasterman, ed., *From Media Hype to Twitter Storm: News Explosions and Their Impact on Issues, Crises, and Public Opinion* (Amsterdam: Amsterdam University Press, 2018), doi:10.5117/9789462982178/ch04. Stuart Ritchie, *Science Fictions: How Fraud, Bias, Negligence, and Hype Undermine the Search for Truth* (New York: Metropolitan Books, 2020).

34. Philip Kitcher, *The Advancement of Science: Science without Legend, Objectivity without Illusions* (New York: Oxford University Press, 1993). Ludwig Fahrbach, "Scientific Revolutions and the Explosion of Scientific Evidence," *Synthese* 194: 5039–5072 (2017), doi:10.1007/s11229-016-1193-y.

35. There is some engineering work to exploit zero-point energy; this is inconclusive at best. Garret Moddel and Olga Dmitriyeva, "Extraction of Zero-Point Energy from the Vacuum: Assessment of Stochastic Electrodynamics-Based Approach as Compared to Other Methods," *Atoms* 7: 51 (2019), doi:10.3390/atoms7020051.

36. Robert J. Schadewald, *Worlds of Their Own: A Brief History of Misguided Ideas: Creationism, Flat-Earthism, Energy Scams, and the Velikovsky Affair* (Xlibris, 2008), 45–84.

37. Gary Taubes, *Bad Science: The Short Life and Weird Times of Cold Fusion* (New York: Random House, 1993).

38. The most sympathetic mainstream assessment of cold fusion is Curtis P. Berlinguette, Yet-Ming Chiang, Jeremy N. Munday et al., "Revisiting the Cold Case of Cold Fusion," *Nature* 570: 45–51 (2019), doi:10.1038/s41586-019-1256-6. And yet, there are scientists who remain hopeful. For example, Jean-Paul Biberian, ed. *Cold Fusion: Advances in Condensed Matter Nuclear Science* (Amsterdam: Elsevier, 2020). For a nontechnical skeptical review, see David W. Ball, "Cold Fusion: Thirty Years Later," *Skeptical Inquirer* 43(1): 28–35 (2019).

39. Sometimes cold fusion is part of a free energy scam. Sadri Hassani, "Why E-Cat is a Hoax," *Skeptical Inquirer* 43(1): 36–40 (2019); Ian Bryce, "Rossi's E-Cat: Exposé Of A Claimed Cold Fusion Device," *Skeptical Inquirer* 43(3): 40–48 (2019).

40. Richard Noakes, *Physics and Psychics: The Occult and the Sciences in Modern Britain* (Cambridge: Cambridge University Press, 2019).

41. Parapsychology seems caught in a pattern of producing research that appears convincing by the standards of the behavioral sciences, which evaporates on further scrutiny. For example, in the past decade, skeptics were challenged by the research of Daryl Bem, then a highly reputable psychologist, that supported psychic powers, particularly in papers such as Daryl J. Bem, "Feeling the Future: Experimental Evidence for Anomalous Retroactive Influences on Cognition and Affect," *Journal of Personality and Social Psychology* 100(3): 407–425 (2011), doi:10.1037/a0021524. It recently became clear that Bem was engaging in multiple forms of malpractice: Susan Blackmore, "Daryl Bem and Psi in the Ganzfeld," *Skeptical Inquirer* 42(1): 44–45 (2018); Susan Blackmore, "Another Scandal for Psychology: Daryl Bem's Data Massage," *Skeptical Inquirer* 43(6): 5 (2019). Parapsychology is

also caught up in psychology's replication crisis. My view is that institutionally, parapsychology occupies a gray area between real and fake science. Taner Edis, "From Creationism to Economics: How Far Should Analyses of Pseudoscience Extend?" *Mètode Science Studies Journal* 8: 141–147 (2018); doi:10.7203/metode.8.10001.

42. I first argued that laboratory parapsychology suffers from a lack of theory and that it does not produce a signal rising above the noise in Taner Edis, *The Ghost in the Universe: God in Light of Modern Science* (Amherst: Prometheus Books, 2002), 192–197. I have seen nothing to change my mind.

43. Brian M. Hughes, *Psychology in Crisis* (London: Palgrave MacMillan, 2018). Note that a replication crisis does not necessarily indicate a preponderance of poor quality studies: Alexander Bird, "Understanding the Replication Crisis as a Base Rate Fallacy," *The British Journal for the Philosophy of Science* (2018), doi:10.1093/bjps/axy051.

44. There are many varieties of physicalism. My sympathies are with the more reductive versions, such as Kevin Morris, *Physicalism Deconstructed: Levels of Reality and the Mind-Body Problem* (New York: Cambridge University Press, 2019); Andrew Melnyk, *A Physicalist Manifesto: Thoroughly Modern Materialism* (New York: Cambridge University Press, 2003).

45. William A. Dembski, *The Design Inference: Eliminating Chance Through Small Probabilities* (New York: Cambridge University Press, 1998); William A. Dembski, *No Free Lunch: Why Specified Complexity Cannot Be Purchased Without Intelligence* (Lanham: Rowman & Littlefield, 2002). For critiques, see Matt Young and Taner Edis, eds., *Why Intelligent Design Fails: A Scientific Critique Of The New Creationism* (New Brunswick: Rutgers University Press, 2004); Olle Häggström, "Intelligent Design and the NFL Theorems," *Biology and Philosophy* 22(2): 217–230 (2007), doi:10.1007/s10539-006-9040-z. Some echoes of the earlier mathematical approach to ID survive; for example, Steinar Thorvaldsen and Ola Hössjer, "Using Statistical Methods to Model the Fine-Tuning of Molecular Machines and Systems," *Journal of Theoretical Biology* 501: 110352 (2020), doi:10.1016/j.jtbi.2020.110352. It is still not likely to produce anything of scientific substance.

46. Taner Edis, "Chance and Necessity—and Intelligent Design?" in Matt Young and Taner Edis, eds., *Why Intelligent Design Fails: A Scientific Critique Of The New Creationism* (New Brunswick: Rutgers University Press, 2004).

47. Martin Davis, "The Myth of Hypercomputation," in Christof Teuscher, ed., *Alan Turing: Life and Legacy of a Great Thinker* (New York: Springer, 2004). The most serious case for hypercomputation I have run into is Selmer Bringsjord and Michael Zenzen, *Superminds: People Harness Hypercompu-*

tation, and More (Dordrecht: Kluwer Academic Publishers, 2003), which strikes me as comparable to parapsychology in its implausibility.

48. Taner Edis and Maarten Boudry, "Beyond Physics? On the Prospects of Finding a Meaningful Oracle," *Foundations of Science* 19(4): 403–422 (2014), doi:10.1007/s106 99-014-9349-z.

49. Chance-and-necessity physicalism only indirectly addresses the traditional argument against physicalism: that minds and consciousness cannot be physical. Progress in explaining consciousness in mainstream biological and neuroscientific terms makes all varieties of physicalism more plausible. Todd E. Feinberg and Jon M. Mallatt, *Consciousness Demystified* (Cambridge: The MIT Press, 2018); Michael S.A. Graziano, *Rethinking Consciousness: A Scientific Theory of Subjective Experience* (New York: W.W. Norton, 2019).

50. Richard A. Muller, *Now: The Physics of Time* (New York: W.W. Norton, 2016), 266–274.

51. Dualist philosophies are most congenial to paranormal beliefs. Jonathan J. Loose, Angus J.L. Menuge, and J.P. Moreland, *The Blackwell Companion to Substance Dualism* (Hoboken: John Wiley and Sons, 2018). Without committing to dualism, anti-physicalist philosophy can also come close to supporting intelligent design; for example, Thomas Nagel, *Mind and Cosmos: Why the Materialist Neo-Darwinian Conception of Nature Is Almost Certainly False* (Oxford: Oxford University Press, 2012). While I am not remotely convinced by such views, I have to acknowledge that they are respectable, though not common, within philosophy.

52. Sabine Hossenfelder, "Science Needs Reason to be Trusted," *Nature Physics* 13: 316–317 (2017), doi:10.1038/nphys4079.

53. Barbara Ryden, *Introduction to Cosmology* (2nd edition, Cambridge: Cambridge University Press, 2017).

54. Simon Friederich, *Multiverse Theories: A Philosophical Perspective* (Cambridge: Cambridge University Press, 2021).

55. See contributions in Radin Dardashti, Richard Dawid, and Karim Thébault, eds., *Why Trust a Theory? Epistemology of Fundamental Physics* (Cambridge: Cambridge University Press, 2019).

56. The most notorious recent example is Lawrence Krauss, *A Universe From Nothing: Why There Is Something Rather than Nothing* (New York: Free Press, 2012). While few would consider Krauss to have succeeded in his grandest ambitions, I think there is some merit to the idea of drawing on physics to more precisely define "nothing" and see what that implies.

Chapter 3

1. David L. Weddle, *Miracles: Wonder and Meaning in World Religions* (New York: New York University Press, 2010). Darryl Caterine and John W. Morehead, eds., *The Paranormal and Popular Culture: A Postmodern Religious Landscape* (New York: Routledge, 2019).

2. Kirsten W. Endres and Andrea Lauser, eds., *Engaging the Spirit World: Popular Beliefs and Practices in Modern Southeast Asia* (New York: Berghahn Books, 2011).

3. Elaine Howard Ecklund and Christopher P. Scheitle, *Religion vs. Science: What Religious People Really Think* (New York: Oxford University Press, 2018), 56.

4. Taner Edis, *The Ghost in the Universe: God in Light of Modern Science* (Amherst: Prometheus Books, 2002). Taner Edis, *Science and Nonbelief* (Westport: Greenwood Press, 2006).

5. Such a view has become associated with "new atheist" writers, such as Sam Harris, *The End of Faith: Religion, Terror, and the Future of Reason* (New York: W.W. Norton, 2005). More serious philosophical approaches are considerably more complicated. Terence Penelhum, *Reason and Religious Faith* (Boulder: Westview Press, 1995).

6. Lara Buchak, "Reason and Faith," in William J. Abraham and Frederick D. Aquino, eds., *The Oxford Handbook of the Epistemology of Theology* (Oxford: Oxford University Press, 2017), 46–63. Steve Wilkens, ed., *Faith and Reason: Three Views* (Downers Grove: IVP Academic, 2014).

7. Ilkka Pyysiäinen, *Magic, Miracles, and Religion: A Scientist's Perspective* (Walnut Creek: AltaMira, 2004). The overwhelming majority of people adopt their family religious beliefs; while humans have a common susceptibility to supernatural beliefs, commitment to particular traditions is largely a matter of instilled loyalty. Benjamin Beit-Hallahmi, *Psychologial Perspectives on Religion and Religiosity* (New York: Routledge, 2015), chapter 3.

8. Lawrence Principe, *The Secrets of Alchemy* (Chicago: University of Chicago Press, 2013). Jan-Olav Henriksen, *Religion as Orientation and Transformation: A Maximalist Theory* (Tübingen: Mohr Siebeck, 2017).

9. David L. Weddle, *Miracles: Wonder and Meaning in World Religions* (New York: New York University Press, 2010). For a defense of religious miracles and mysticism in the context of paranormal claims, see Paul Marshall, "Mystical Experiences as Windows on Reality," in Edward F. Kelly, Adam Crabtree, and Paul Marshall, eds., *Beyond Physicalism: Toward Reconciliation of Science and Spirituality* (Lanham: Rowman & Littlefield, 2015). However, by the same token, criticisms of the paranormal also apply to

miracles and mysticism.

10. James A. Van Slyke, *The Cognitive Science of Religion* (Burlington: Ashgate Publishing, 2011), chapter 3.

11. Lee McIntyre, *The Scientific Attitude: Defending Science from Denial, Fraud, and Pseudoscience* (Cambridge: The MIT Press, 2019).

12. This recalls the separate spheres or "non-overlapping magisteria" famously defended by Stephen Jay Gould, *Rocks of Ages: Science and Religion in the Fullness of Life* (New York: Ballantine Publishing, 1999). Believers in paranormal realities, however, do not usually accept such a division of labor; Egil Asprem, *The Problem of Disenchantment: Scientific Naturalism and Esoteric Discourse 1900–1939* (Leiden: Brill, 2014). Christians are not always happy with it either: Gary N. Fugle, *Laying Down Arms to Heal the Creation-Evolution Divide* (Eugene: Wipf and Stock, 2015), 83–88.

13. Emily G.Liquin, S. Emlen Metz, and Tania Lombrozo, "Science Demands Explanation, Religion Tolerates Mystery," *Cognition* 204: 104398 (2020), doi:10.1016/j.cognition.2020.104398. Evasiveness is, unfortunately, often described as "unfalsifiability"; for example, Justin P. Friesen, Troy H. Campbell, and Aaron C. Kay, "The Psychological Advantage of Unfalsifiability: The Appeal of Untestable Religious and Political Ideologies," *Journal of Personality and Social Psychology* 108(3): 515–529 (2015), doi:10.1037/pspp0000018.

14. The notorious "experimenter effect." See, for example, Richard Wiseman and Marilyn Schlitz, "Experimenter Effects and the Remote Detection of Staring," *Journal of Parapsychology* 61(3): 197–208 (1997); David E.H. Jones, *Why Are We Conscious? A Scientist's Take on Consciousness and Extrasensory Perception* (Singapore: Pan Stanford Publishing, 2017), chapter 6.

15. Christopher C. French and Anna Stone, *Anomalistic Psychology: Exploring Paranormal Belief and Experience* (Basingstoke: Palgrave MacMillan, 2014), 39–40. Gordon Pennycook et. al, "Analytic Cognitive Style Predicts Religious and Paranormal Belief," *Cognition* 123(3): 335–346 (2012), doi:10.1016/j.cognition.2012.03.003. For criticism of overly simplistic dual-process accounts, see Olli-Pekka Vainio, "What Does Theology Have to Do With Religion? Dual-process Accounts, Cognitive Science of Religion and a Curious Blind Spot in Contemporary Theorizing," *Open Theology* 2: 106–112 (2016), doi:10.1515/opth-2016-0008.

16. Mary Jo Nye, "N-Rays: An Episode in the History and Psychology of Science," *Historical Studies in the Physical Sciences* 11(1): 125–156 (1980), doi:10.2307/27757473. Monwhea Jeng, "A Selected History of Expectation Bias in Physics," *American Journal of Physics* 74(7): 578–583 (2006), doi:10.1119/1.2186333.

17. David B. Yaden et al., "The Noetic Quality: A Multimethod Exploratory Study," *Psychology of Consciousness: Theory, Research, and Practice* 4(1): 54–62 (2017), doi:10.1037/cns0000098.

18. Joanna Maselko, "The Neurophysiology of Religious Experience," in Kenneth I. Pargament, Julie J. Exline, and James W. Jones, eds., *APA Handbook of Psychology, Religion, and Spirituality (Vol. 1): Context, Theory, And Research* (Washington: American Psychological Association, 2013), doi:10.1037/14045-011. Michael J. Winkelman, "The Mechanisms of Psychedelic Visionary Experiences: Hypotheses from Evolutionary Psychology," *Frontiers in Neuroscience* 11: 539 (2017), doi:10.3389/fnins.2017.00539. Defenders of mystical ways of knowing are apt to argue that neuroscience does not address the question of whether mystical experiences are windows into supernatural realities; Richard H. Jones, "Limitations on the Scientific Study of Drug-Enabled Mystical Experiences," *Zygon* 54(3): 756–792 (2019), doi:10.1111/zygo.12546. That conclusion is excessive; neuroscience tends to naturalize and therefore deflate mystical claims.

19. Scholars of mysticism debate between social constructivism and the claim that there is a common core. Both seem to be correct to some degree. Torben Hammersholt, "Steven T. Katz's Philosophy of Mysticism Revisited," *Journal of the American Academy of Religion* 81(2): 467–490 (2013), doi:10.1093/jaarel/lft001. Ralph W. Hood Jr., "The Common Core Thesis in the Study of Mysticism," *Oxford Research Encyclopedia of Religion* (2016), doi:10.1093/acrefore/9780199340378.013.241.

20. Joseph O. Baker and Scott Draper, "Diverse Supernatural Portfolios: Certitude, Exclusivity, and the Curvilinear Relationship Between Religiosity and Paranormal Beliefs," *Journal for the Scientific Study of Religion* 49: 413–424 (2010), doi:10.1111/j.1468-5906.2010.01519.x. Marc Stewart Wilson, Joseph Bulbulia, and Chris G. Sibley, "Differences and Similarities in Religious and Paranormal Beliefs: A Typology of Distinct Faith Signatures," *Religion, Brain & Behavior* 4(2): 104–126 (2014), doi:10.1080/21535 99X.2013.779934.

21. Marc Andersen, Uffe Schjoedt, Kristoffer L. Nielbo, and Jesper Sørensen, "Mystical Experience in the Lab," *Method & Theory in the Study of Religion* 26(3): 217–245 (2014), doi:10.1163/15700682-12341323. Irene Cristofori et. al, "Neural Correlates of Mystical Experience," *Neuropsychologia* 80: 212–220 (2016), doi:10.1016/j.neuropsychologia.2015.11.021. Some philosophers think that mystical truth claims can be reconciled with scientific analyses; Richard H. Jones, *Philosophy of Mysticism: Raids on the Ineffable* (Albany: State University of New York Press, 2016); I am not convinced.

22. The theory-ladenness of data, a commonplace in the philosophy of science,

is sometimes taken to represent a challenge to the rationality of science. Samuel Schindler, "Theory-laden Experimentation," *Studies in History and Philosophy of Science Part A* 44(1): 89–101 (2013), doi:10.1016/j.shpsa.2012.07.010. With claims of unmediated paranormal experience, recognizing theory-laden "data" is important for science.

23. Joe Nickell, *The Science of Miracles: Investigating the Incredible* (Amherst: Prometheus, 2013). Larry Shapiro, *The Miracle Myth: Why Belief in the Resurrection and the Supernatural Is Unjustified* (New York: Columbia University Press, 2016).

24. I am ignoring all technical issues having to do with black hole information loss problems. Donald Marolf, "The Black Hole Information Problem: Past, Present, and Future," *Reports on Progress in Physics* 80(9): 092001 (2017), doi:10.1088/1361-6633/aa77cc. Even with no loss, anything thrown into a black hole is lost for all practical purposes.

25. Taner Edis, "Cosmic Conspiracy Theories: How Theologies Evade Science," in Joseph Seckbach and Richard Gordon, eds., *Theology and Science: From Genesis to Astrobiology* (Singapore: World Scientific, 2019).

26. C. Mackenzie Brown, ed., *Asian Religious Responses to Darwinism: Evolutionary Theories in Middle Eastern, South Asian, and East Asian Cultural Contexts* (Cham: Springer Nature, 2020).

27. Some form of compatibility and compromise is what most defenders of evolution hope for. Eugenie C. Scott, *Evolution vs. Creationism: An Introduction* (2nd edition, Berkeley: University of California Press, 2009). Still, I can't help but have some sympathy for creationists who reject such compromises. Christopher Gieschen, *Is Evolution Compatible with Christianity?* (Eugene: Wipf & Stock Publishers, 2019).

28. Edward J. Larson, *Evolution: The Remarkable History of a Scientific Theory* (New York: Modern Library, 2004). James R. Moore, ed., *History, Humanity, and Evolution: Essays for John C. Greene* (New York: Cambridge University Press, 1989).

29. Alexander Werth, "Avoiding the Pitfall of Progress and Associated Perils of Evolutionary Education," *Evolution: Education and Outreach* 5: 249–265 (2012), doi:10.1007/s12052-012-0417-y. Arthur O. Lovejoy, *The Great Chain of Being: A Study of the History of an Idea* (Cambridge: Harvard University Press, 1936).

30. While evolutionary mechanisms lack purpose, there are still interesting philosophical questions concerning the appearance of teleology. Daniel W. McShea, "Hierarchy: The Source of Teleology in Evolution," in Niles Eldredge, Telmo Pievani, Emanuele Serrelli, and Ilya Temkin, eds., *Evolutionary Theory: A Hierarchical Perspective* (Chicago: University of Chi-

cago Press, 2016), 86–102. Michael Ruse, "Evolutionary Biology and the Question of Teleology," *Studies in History and Philosophy of Science Part C: Studies in History and Philosophy of Biological and Biomedical Sciences* 58: 100–106 (2016), doi:/10.1016/j.shpsc.2015.12.001.

31. For example, John F. Haught, *Deeper than Darwin: The Prospect of Religion in the Age of Evolution* (Boulder: Westview Press, 2003). Even when making a theological argument for autonomy, Haught can't help slipping into thinking of evolution as a teleological process.

32. Taner Edis and Maarten Boudry, "Beyond Physics? On the Prospects of Finding a Meaningful Oracle," *Foundations of Science* 19(4): 403–422 (2014), doi:10.1007/s106 99-014-9349-z. Taner Edis, *Science and Nonbelief* (Westport: Greenwood Press, 2006).

33. Arne Dietrich, *How Creativity Happens in the Brain* (New York: Palgrave MacMillan, 2015). Maria Kronfeldner, *Darwinian Creativity and Memetics* (Durham: Acumen, 2011) has a more critical view, but *some* role for variation and selection in providing a nonmagical form of creativity seems secure.

34. Jonathan H. Turner, Alexandra Maryanski, Anders Klostergaard Petersen, and Armin W. Geertz, eds., *The Emergence and Evolution of Religion: By Means of Natural Selection* (Milton Park: Taylor & Francis, 2017). Hans Van Eyghen and Konrad Szocik, *Revising Cognitive and Evolutionary Science of Religion: Religion as an Adaptation* (Cham: Springer Nature, 2021).

35. Taner Edis, "Cosmic Conspiracy Theories: How Theologies Evade Science," in Joseph Seckbach and Richard Gordon, eds., *Theology and Science: From Genesis to Astrobiology* (Singapore: World Scientific, 2019).

36. Classic examples are J. Allen Hynek, *The UFO Experience: A Scientific Inquiry* (Chicago: Henry Regnery Company, 1972), Peter A. Sturrock, *The UFO Enigma: A New Review of the Physical Evidence* (New York: Warner Books, 1999). With public interest in UFOs waning over the last two decades, it is harder to find recent examples of scientists declaring UFOs a mystery.

37. "Ehab Abouheif & Taner Edis on Evolution and Islam" (2009), https://www.youtube.com/watch?v=RS7mPtsfejg (accessed February 27, 2021).

38. Jamie Carlin Watson, *Expertise: A Philosophical Introduction* (London: Bloomsbury Academic, 2021).

39. "Evolution, Creationism, Intelligent Design," *Gallup*, https://news.gallup.com/poll/21814/Evolution-Creationism-Intelligent-Design.aspx, accessed December 31, 2020. Adam Laats, *Creationism USA: Bridging the Impasse on Teaching Evolution* (New York: Oxford University Press, 2021), 24–27. Kevin McCain and Kostas Kampourakis, "Which Question Do Polls about

Evolution and Belief Really Ask, and Why Does It Matter?" *Public Understanding of Science* 27(1)): 2–10 (2018), doi:10.1177/0963662516642726.

40. James Bell et al., *The World's Muslims: Religion, Politics and Society* (Washington: Pew Research Center, 2013), 132–133.

41. Elaine Howard Ecklund et al., *Secularity and Science: What Scientists around the World Really Think about Religion* (New York: Oxford University Press, 2019). George M. Marsden, "Religious Discrimination in Academia" *Society* 52: 19–22 (2015), doi:10.1007/s12115-014-9853-3.

42. As more applied scientists are surveyed, religiosity increases: for example, Elaine Howard Ecklund and Christopher P. Scheitle, *Religion vs. Science: What Religious People Really Think* (New York: Oxford University Press, 2018), 56–62. On the differing mindset of applied scientists, particularly engineers, see evidence summarized in Diego Gambetta and Steffen Hertog, "Why Are There So Many Engineers Among Islamic Radicals?" *European Journal of Sociology* 50(2): 201–230 (2009), doi:10.1017/S0003975609990129; Diego Gambetta and Steffen Hertog, *Engineers of Jihad: The Curious Connection between Violent Extremism and Education* (Princeton: Princeton University Press, 2016), chapter 6.

43. Elaine Howard Ecklund et al., *Secularity and Science: What Scientists around the World Really Think about Religion* (New York: Oxford University Press, 2019). This observation might generalize to broader populations as well; Magali Clobert and Vassilis Saroglou, "Religion, Paranormal Beliefs, and Distrust in Science: Comparing East Versus West," *Archive for the Psychology of Religion* 37(2): 185–199 (2015), doi:10.1163/15736121-12341302.

44. Taner Edis, *Islam Evolving: Radicalism, Reformation, and the Uneasy Relationship with the Secular West* (Amherst: Prometheus, 2016). Taner Edis, "The Turkish Model of Islamic Creationism," *Almagest* 12: 40–65 (2021), doi:10.1484/J.ALMAGEST.5.125385.

45. Self-selection has been most studied in the context of the political liberalism of faculty, but similar arguments plausibly apply to science students and religious liberalism as well. Neil Gross, *Why Are Professors Liberal and Why Do Conservatives Care?* (Cambridge: Harvard University Press, 2013). Some relevant research is surveyed in Benjamin Beit-Hallahmi, *Psychological Perspectives on Religion and Religiosity* (New York: Routledge, 2015), 73–75.

46. Georgia L. Irby-Massie and Paul T. Keyser, eds., *Greek Science of the Hellenistic Era: A Sourcebook* (New York: Routledge, 2002). J. Edward Wright, *The Early History of Heaven* (New York: Oxford University Press, 2000).

47. Taner Edis, *An Illusion of Harmony: Science and Religion in Islam* (Amherst: Prometheus Books, 2007), 33–52. Peter E. Pormann and Emilie Sav-

age-Smith, *Medieval Islamic Medicine* (Edinburgh: Edinburgh University Press, 2007).

48. William R. Newman, *Newton the Alchemist: Science, Enigma, and the Quest for Nature's "Secret Fire"* (Princeton: Princeton University Press, 2018). Rob Iliffe, *Priest of Nature: The Religious Worlds of Isaac Newton* (New York: Oxford University Press, 2017).

49. Richard Noakes, *Physics and Psychics: The Occult and the Sciences in Modern Britain* (Cambridge: Cambridge University Press, 2019).

50. Yaakov Malkin, *Judaism Without God: Judaism as Culture, Bible as Literature* (Piscataway: Gorgias Press, 2009). Lloyd Geering, *Christianity Without God* (Santa Rosa: Polebridge Press, 2002). Stephen Batchelor, *After Buddhism: Rethinking the Dharma for a Secular Age* (New Haven: Yale University Press, 2015).

51. Gregory W. Dawes, *Galileo and the Conflict Between Religion and Science* (New York: Routledge, 2016).

52. Ian Barbour's typology of conflict, independence, dialogue, and integration has been influential, especially in introductory texts; for example, Alister E. McGrath, *Science & Religion: A New Introduction* (3rd edition, Hoboken: Wiley Blackwell, 2020). To reflect what my students tend to come up with, I emphasize harmony, separation, and conflict instead. Since I think all such models are inadequate, including Barbour's, I don't think the typology adopted is very important.

53. Taner Edis, *Islam Evolving: Radicalism, Reformation, and the Uneasy Relationship with the Secular West* (Amherst: Prometheus, 2016).

54. Robert T. Pennock, "Can't Philosophers Tell the Difference Between Science and Religion?: Demarcation Revisited," *Synthese* 178: 177–206 (2011), doi:10.1007/s11229-009-9547-3. Taner Edis, "Atheism and the Rise of Science," in Stephen Bullivant and Michael Ruse, eds., *The Oxford Handbook of Atheism* (Oxford: Oxford University Press, 2013).

55. Jeff Hardin, Ronald L. Numbers, and Ronald A. Binzley, eds., *The Warfare Between Science and Religion: The Idea that Wouldn't Die* (Baltimore: Johns Hopkins University Press, 2018).

Chapter 4

1. I looked at the CSI Fellows and Scientific and Technical Consultants listed on the cover of *Skeptical Inquirer* 44(6), 2020, and counted their academic backgrounds. Physics and astronomy had the largest share, at 19%, followed by psychologists at 17%. In contrast, those in medicine came in at 9% and philosophy at 5%. These numbers are only a rough indication, but

my impression is that physicists and psychologists are most heavily represented among my books critical of paranormal beliefs and in the pages of the *Skeptical Inquirer.*

2. This seems to be one of the motivations for some textbooks I used during the first years I taught Weird Science. Charles M. Wynn and Arthur W. Wiggins, *Quantum Leaps in the Wrong Direction: Where Real Science Ends—And Pseudoscience Begins* (2nd edition, New York: Oxford University Press, 2017). Theodore Schick, Jr. & Lewis Vaughn, *How to Think About Weird Things: Critical Thinking for a New Age* (8th edition, New York: Mc-Graw-Hill Education, 2019)

3. Maarten Boudry and Massimo Pigliucci, eds., *Science Unlimited? The Challenges of Scientism* (Chicago: University of Chicago Press, 2018). Richard N. Williams and Daniel N. Robinson, eds., *Scientism: The New Orthodoxy* (New York: Bloomsbury Academic, 2015). Johan Hietanen et al., "How *Not* to Criticise Scientism," *Metaphilosophy* 51: 522–547 (2020), doi:10.1111/meta.12443. Some of this chapter draws on my contribution to the volume edited by Boudry and Pigliucci, "Two Cheers for Scientism."

4. For example, Lawrence Krauss in an interview by Ross Anderson, "Has Physics Made Philosophy and Religion Obsolete?" *Atlantic* April 23, 2012, https://www.theatlantic.com/technology/archive/2012/04/has-physics-made-philosophy-and-religion-obsolete/256203/, accessed January 2, 2021. Also, Massimo Pigliucci, "Neil deGrasse Tyson and the Value of Philosophy," *Huffpost* May 16, 2014, https://www.huffpost.com/entry/neil-degrasse-tyson-and-the-value-of-philosophy_b_5330216, accessed January 2, 2021.

5. Mary B. Hesse, *Forces and Fields: The Concept of Action at a Distance in the History of Physics* (London: T. Nelson, 1961).

6. These are known as gauge symmetries, and local forms of gauge symmetries are fundamental to our current understanding of forces. Chris Quigg, *Gauge Theories of the Strong, Weak, and Electromagnetic Interactions* (2nd edition, Princeton: Princeton University Press, 2013).

7. For example, Wigner functions, which are real functions of position and momentum, rather the more commonly used wave functions which are complex functions of position. Thomas L. Curtright, David B. Fairlie, and Cosmas K. Zachos, *A Concise Treatise on Quantum Mechanics in Phase Space* (Singapore: World Scientific, 2014).

8. It may appear that I am endorsing a kind of ontological relativism, according to which objects do not exist in any mind-dependent sense. No; I am just observing that our descriptions do not directly map onto mind-independent realities that constrain our descriptions. Philosophers continue to

be interested in disputes over realism and relativism; for example, Dominik Finkelde and Paul M. Livingston, eds., *Idealism, Relativism, and Realism: New Essays on Objectivity Beyond the Analytic-Continental Divide* (Berlin: Walter de Gruyter GmbH, 2020). I don't find such debates helpful for questions that interest me.

9. On quantum randomness, see Karl Svozil, *Physical (A)Causality: Determinism, Randomness and Uncaused Events* (Cham: Springer Nature, 2018). Bohmian hidden-variable interpretations of quantum mechanics restore an appearance of determinism to physics while keeping outcomes random; they are equivalents of my gremlins. There is nothing empirically wrong with Bohmian approaches, but they can be misleading. Indeed, David Bohm's views are often used to support paranormal claims; for example, Alan Ross Hugenot, *The New Science of Consciousness Survival and the Metaparadigm Shift to a Conscious Universe* (Indianapolis: Dog Ear Publishing, 2016). Bohm's writings encouraged at least some level of mysticism; David Bohm, *Wholeness and the Implicate Order* (London: Ark, 1983).

10. Michael Martin and Ricki Monnier, eds., *The Impossibility of God* (Amherst: Prometheus Books, 2003), has some contributions that explore such paradoxes.

11. Observations about the disconnection between religious studies and the philosophy of religion are common; for example, Thomas A. Lewis, *Why Philosophy Matters for the Study of Religion—and Vice Versa* (New York: Oxford University Press, 2015); Kevin Schilbrak, *Philosophy and the Study of Religions: A Manifesto* (Malden: Wiley-Blackwell, 2014). Some philosophers of religion recognize the perpetual stalemate of their philosophical tradition; Graham Oppy, *Reinventing Philosophy of Religion: An Opinionated Introduction* (New York: Palgrave Macmillan 2014). In effect, Oppy embraces the stalemate. I think that those of us interested in serious knowledge claims should instead stop playing the game called "philosophy of religion." Taner Edis, "Doubt and Submission: Why Evil Is a Minor Problem for Islam," in John W. Loftus, ed. *The Incompatibility of God and Horrendous Suffering* (Denver: Global Center for Religious Research, 2021).

12. This is the territory of metaphilosophy; Nicholas Rescher, *Metaphilosophy: Philosophy in Philosophical Perspective* (Lanham: Lexington Books, 2014). Though not an explicitly metaphilosophical text, one of my favorite examples of gentle criticisms of philosophy from within is Simon Blackburn, *Truth: A Guide* (New York: Oxford University Press, 2005), since I found it such a pleasure to read.

13. Max Tegmark appears to think that most theoretical physicists are Pla-

tonists, and defends a variety of Platonism: Max Tegmark, *Our Mathematical Universe: My Quest for the Ultimate Nature of Reality* (New York: Alfred E. Knopf, 2014). I'm not so sure about Platonism among theorists; I doubt most care. I'm fairly sure that Tegmark's Platonism is untethered from reality. For another variety of Platonism, see Roger Penrose, "Mathematics, the Mind, and the Physical World," in John Polkinghorne, ed., *Meaning in Mathematics* (Oxford: Oxford University Press, 2011).

14. James Robert Brown, *Platonism, Naturalism, and Mathematical Knowledge* (New York: Routledge, 2012), defends an anti-naturalist, Platonist "seeing with the mind's eye," but this, I think, is still a sophisticated version of mystical illumination.

15. Such an approach is sometimes termed "philosophy of mathematical practice," but in many ways it is more promising than traditional debates over the metaphysics of mathematical objects. Roi Wagner, *Making and Breaking Mathematical Sense: Histories and Philosophies of Mathematical Practice* (Princeton: Princeton University Press, 2017).

16. Morris Kline, *Mathematics: The Loss of Certainty* (New York: Fall River Press, 1980).

17. Jody Azzouni, *Talking about Nothing: Numbers, Hallucinations, and Fictions* (Oxford: Oxford University Press, 2010).

18. George Lakoff and Rafael E. Núñez, *Where Mathematics Comes From: How the Embodied Mind Brings Mathematics into Being* (New York: Basic Books, 2000). A cognitive approach also illuminates the use of math in physics. Ryan D. Tweny, "Metaphor and Model-Based Reasoning in Maxwell's Mathematical Physics," in Lorenzo Magnani, ed., *Model-Based Reasoning in Science and Technology: Theoretical and Cognitive Issues* (Berlin: Springer-Verlag, 2014).

19. Paul Ernest, *Social Constructivism as a Philosophy of Mathematics* (Albany: State University of New York Press, 1998). Reuben Hersh, *What Is Mathematics, Really?* (New York: Oxford University Press, 1997), is still a good introduction to such "humanist" philosophies of mathematics. Notably, non-Platonic approaches to math continue to be most engaged in questions concerning mathematics education.

20. Henry Plotkin, *The Imagined World Made Real: Towards a Natural Science of Culture* (New Brunswick: Rutgers University Press, 2003), though somewhat dated, still describes the agenda pretty well. Pascal Boyer, *Minds Make Societies: How Cognition Explains the World Humans Create* (New Haven: Yale University Press, 2018), presents a more recent view, even suggesting that the naïve concept of "culture" may not be that useful any more.

21. See contributions in Giovanni Boniolo, Majda Trobok, and Paolo Budinich,

eds., *The Role of Mathematics in Physical Sciences: Interdisciplinary and Philosophical Aspects* (Dordrecht: Springer, 2005).

22. Claude E. Shannon and Warren Weaver, *The Mathematical Theory of Communication* (Urbana: University of Illinois Press, 1949).

23. There is now a complex field of philosophy of engineering, addressing such social and ethical questions and much more. Anthonie W.M. Meijers, ed., *Philosophy of Technology and Engineering Sciences* (Amsterdam: Elsevier, 2009). Diane P. Michelfelder and Neelke Doorn, eds., *The Routledge Handbook of the Philosophy of Engineering* (New York: Routledge, 2021).

24. William Bynum, *The History of Medicine: A Very Short Introduction* (Oxford: Oxford University Press, 2008). Roberta Bivins, *Alternative Medicine?: A History* (Oxford: Oxford University Press, 2007).

25. Stephen Barrett, "Dubious Aspects of Osteopathy," *Quackwatch*, February 11, 2018, https://quackwatch.org/consumer-education/qa/osteo/, accessed January 3, 2021.

26. Brent A. Bauer, ed., *Mayo Clinic Guide to Integrative Medicine: Conventional Remedies Meet Alternative Therapies to Transform Health* (New York: Time Incorporated Books, 2017). Bonnie McLean, *Integrative Medicine: The Return of the Soul to Healthcare* (Bloomington: Balboa Press, 2015).

27. David H. Gorski, "'Integrative' Medicine: Integrating Quackery with Science-Based Medicine," in Allison B. Kaufman and James C. Kaufman, eds., *Pseudoscience: The Conspiracy Against Science* (Cambridge: The MIT Press, 2018). Eugenie V. Mielczarek and Brian D. Engler, "Selling Pseudoscience: A Rent in the Fabric of American Medicine," *Skeptical Inquirer* 38(3): 44–51 (2014).

28. Edzard Ernst, Max H. Pittler, Clare Stevinson, and Adrian White, eds., *The Desktop Guide to Complementary and Alternative Medicine: An Evidence-Based Approach* (London: Harcourt, 2001). David H. Gorski and Steven P. Novella, "Clinical Trials of Integrative Medicine: Testing Whether Magic Works?" *Trends in Molecular Medicine* 20 (9): 473–76 (2014), doi:10.1016/j.molmed.2014.06.007.

29. For example, Andrew Miles, "On a Medicine of the Whole Person: Away from Scientist Reductionism and Towards the Embrace of the Complex in Clinical Practice," *Journal of Evaluation in Clinical Practice* 15: 941–49 (2009), doi:10.1111/j.1365-2753.2009.01354.x. Such objections, with their accusations of "microfascism," are overblown. Valid criticisms can be incorporated into the practice of evidence-based medicine: Marie-Caroline Schulte, *Evidence-Based Medicine—A Paradigm Ready To Be Challenged?: How Scientific Evidence Shapes Our Understanding And Use Of Medicine* (Berlin: J.B. Metzler, 2020), doi:10.1007/978-3-476-05703-7.

30. I give examples of such charges in Taner Edis, "Two Cheers for Scientism," in Maarten Boudry and Massimo Pigliucci, eds., *Science Unlimited? The Challenges of Scientism* (Chicago: University of Chicago Press, 2018).

31. David M. Steinhorn, Jana Din, and Angela Johnson, "Healing, Spirituality and Integrative Medicine," *Annals of Palliative Medicine* 6(3) (2017), doi:10.21037/apm.2017.05.01. Religious and spiritual beliefs often prompt use of alternative medicine; Christopher G. Ellison, Matt Bradshaw, and Cheryl A. Roberts, "Spiritual and Religious Identities Predict the Use of Complementary and Alternative Medicine among US Adults," *Preventive Medicine* 54(1): 9–12 (2012), doi:10.1016/j.ypmed.2011.08.029.

32. Lee McIntyre, *Dark Ages: The Case for a Science of Human Behavior* (Cambridge: MIT Press, 2006), 20–21. This assumes that social science has goals of investigation and explanation similar to the natural sciences. Some philosophers and social scientists argue that social science is more concerned with meaning and interpretation; see Alex Rosenberg, *Philosophy of Social Science* (5th edition, Boulder: Westview Press, 2016).

33. Lee C. McIntyre, *Laws And Explanation In The Social Sciences: Defending a Science of Human Behavior* (Boulder: Westview Press, 1998), makes a case for lawlike explanations in social science by analogy with natural sciences studying highly complex systems. There is some progress in this direction: Ton Jörg, *New Thinking in Complexity for the Social Sciences and Humanities: A Generative, Transdisciplinary Approach* (Dordrecht: Springer, 2011); David Byrne and Gill Callaghan, *Complexity Theory and the Social Sciences: The State of the Art* (New York: Routledge, 2014). To me, however, much of this seems to be premature talk of scientific revolutions, overpromising, and getting caught up in slippery metaphors. I agree that complexity is not a barrier in principle to a more rigorous social science, but whether it will continue to remain a practical obstacle will be seen through further developments. Anchoring social science on human cognition seems promising, but not yet developed fully; Pascal Boyer, *Minds Make Societies: How Cognition Explains the World Humans Create* (New Haven: Yale University Press, 2018). For now, a pragmatic philosophy espoused by social scientists seems best, such as Marta Trzebiatowska and Steve Bruce, *Why Are Women More Religious Than Men?* (Oxford: Oxford University Press, 2012), 170–172.

34. There are ways to handle limitations with qualitative data, which often come down to making sure the data used *does* behave like genuine quantities. Validating scales for such data is not, however, trivial, and there's no reason to think it's regularly done properly, given all the other more visible problems with practices in the human sciences. Stuart Ritchie, *Science Fictions: How Fraud, Bias, Negligence, and Hype Undermine the Search for Truth* (New York: Metropolitan Books, 2020). Brian M. Hughes, *Psychology*

in Crisis (London: Palgrave MacMillan, 2018).

35. Ida Kubiszewski et al., "Beyond GDP: Measuring and Achieving Global Genuine Progress," *Ecological Economics* 93: 57–68 (2013), doi:10.1016/j.ecolecon.2013.04.019. Christine Bauhardt and Wendy Harcourt, eds., *Feminist Political Ecology and the Economics of Care: In Search of Economic Alternatives* (New York: Routledge, 2019). Jeff Madrick, *Seven Bad Ideas: How Mainstream Economists Have Damaged America and the World* (New York: Alfred A. Knopf, 2014). Debra Satz, *Why Some Things Should Not be For Sale: The Moral Limits of Markets* (New York: Oxford University Press, 2010). Jonathan Aldred, *Licence to be Bad: How Economics Corrupted Us* (London: Allen Lane, 2019).

36. Or try a better big picture of economic life. For example, Kate Raworth, *Doughnut Economics: Seven Ways to Think Like a 21st Century Economist* (White River Junction: Chelsea Green Publishing, 2017).

37. Mario Bunge, *Finding Philosophy in Social Science* (New Haven: Yale University Press, 1996). Steve Keen, *Debunking Economics: The Naked Emperor Dethroned?* (London: Zed Books, 2011). Such criticism, however, often depends on mid-twentieth century notions of science and pseudoscience. A more modern, and hence more ambiguous view, emerges from the contributions in Don Ross and Harold Kincaid, eds., *The Oxford Handbook of Philosophy of Economics* (Oxford: Oxford University Press, 2009).

38. Ángel Rodriguez, Jorge Turmo, and Oscar Vara, *Financial Crisis and the Failure of Economic Theory* (New York: Routledge, 2016). Richard Bookstaber, *The End of Theory: Financial Crises, the Failure of Economics, and the Sweep of Human Interaction* (Princeton: Princeton University Press, 2017).

39. Philip Mirowski, *Never Let a Serious Crisis Go to Waste: How Neoliberalism Survived the Financial Meltdown* (Brooklyn: Verso, 2013).

40. Joe Earle, Cahal Moran, and Zach Ward-Perkins, *The Econocracy: On the Perils of Leaving Economics to the Experts* (Manchester: Manchester University Press, 2017). Heather Boushey, *Unbound: How Inequality Constricts Our Economy and What We Can Do about It* (Cambridge: Harvard University Press, 2019).

41. Taner Edis, "From Creationism to Economics: How Far Should Analyses of Pseudoscience Extend?" *Mètode Science Studies Journal* 8: 141–147 (2018), doi:10.7203/metode.8.10001.

42. Frank Cioffi, *Freud and the Question of Pseudoscience* (Chicago: Open Court, 1998). Frederick Crews, *Freud: The Making of an Illusion* (New York: Metropolitan Books, 2017).

43. Many philosophers are uncomfortable with the sharp separation of fact and value: Giancarlo Marchetti and Sarin Marchetti, eds., *Facts and Values: The*

Ethics and Metaphysics of Normativity (New York: Routledge, 2016). While I share some of this discomfort, I don't think the approaches represented in this volume are helpful *except* when they bring biologically-rooted values into their picture. Accounts that explicitly foreground biology and neuroscience, such as Patricia S. Churchland, *Braintrust: What Neuroscience Tells Us about Morality* (Princeton: Princeton University Press, 2018), seem much more on track.

44. Helen Small, *The Value of the Humanities* (Oxford: Oxford University Press, 2013).

45. Thomas Nagel, *The View From Nowhere* (New York: Oxford University Press, 1986). I am not generally a fan of Nagel's philosophy, especially in its anti-Darwinian, almost intelligent design-like forms, as in Thomas Nagel, *Mind and Cosmos: Why the Materialist Neo-Darwinian Conception of Nature Is Almost Certainly False* (Oxford: Oxford University Press, 2012). But the "view from nowhere" metaphor is just too good not to appropriate.

46. Carol E. Cleland, "Prediction and Explanation in Historical Natural Science," *The British Journal for the Philosophy of Science* 62(3): 551–582 (2011), doi:10.1093/bjps/axq024.

47. Calvin Schermerhorn, *The Business of Slavery and the Rise of American Capitalism, 1815–1860* (New Haven: Yale University Press, 2015). Sven Beckert and Christine Desan, eds., *American Capitalism: New Histories* (New York: Columbia University Press, 2018).

48. The notion that all historical chronology is wrong is one of the weirdest forms of weirdness I have encountered. Emmet Scott, *Guide to the Phantom Dark Age* (New York: Algora Publishing, 2014). Anatoly Fomenko, *The Issue With Chronology* (Lulu.com, 2016).

49. Out of curiosity, I took a look at Maud Ellmann, *Psychoanalytic Literary Criticism* (New York: Routledge, 2014). I can't say I'm impressed, but literary criticism is so far outside of my concerns that if literary critics find value in psychoanalysis, it's none of my business to object.

50. Robert Boyers, *The Tyranny of Virtue: Identity, the Academy, and the Hunt for Political Heresies* (New York: Scribner, 2019). According to Helen Pluckrose and James Lindsay, *Cynical Theories: How Activist Scholarship Made Everything about Race, Gender, and Identity—and Why This Harms Everyone* (Durham: Pitchstone Publishing, 2020), "applied postmodernism" is more rampant. I'm not entirely convinced by Pluckrose and Lindsay; the genre of exposing leftish craziness to provide a contrast to the virtues of an idealized status quo conservatism doesn't inspire much confidence.

51. It might seem odd that I argue both in favor of physicalism, which inevitably implies a measure of reductionism, and pluralism about the sciences,

which is more often associated with anti-reductionist stances. Done correctly, however, reductionism can help us understand a more pluralist landscape of knowledge. Christian Sachse, *Reductionism in the Philosophy of Science* (Frankfurt: Ontos-Verlag, 2013). Stéphanie Ruphy, *Scientific Pluralism Reconsidered: A New Approach to the (Dis)Unity of Science* (Pittsburgh: University of Pittsburgh Press, 2016).

52. This is the issue of "multiple realization," most often discussed in the context of neuroscience and the philosophy of mind. For example, Thomas W. Polger and Lawrence A. Shapiro, *The Multiple Realization Book* (New York: Oxford University Press, 2016).

53. Setting $dR/dt = c$, the speed of light, and solving for the time when this happens, we get $t = 3\tau \ln(3\tau c) + \tau \ln(4\pi/3N_0 V_1) = 3400$ years, with an initial population of $N_0 = 8$ billion and the absurd assumption that all each person needs to survive is the resources in a prison cell with a volume of $V_1 = 8$ cubic meters. The time scale $\tau = 33.83$ years is for exponential growth of 3% a year.

54. Taner Edis and Amy S. Bix, "Flights of Fancy: The '1001 Inventions' Exhibition and Popular Misrepresentations of Medieval Muslim Science and Technology," in Sonja Brentjes, Taner Edis, and Lutz Richter-Bernburg, eds., *1001 Distortions: How (Not) to Narrate History of Science, Medicine, and Technology in Non-Western Cultures* (Würtzburg: Ergon-Verlag, 2016).

55. Stefaan Blancke, Maarten Boudry, and Massimo Pigliucci, "Why do Irrational Beliefs Mimic Science? The Cultural Evolution of Pseudoscience," *Theoria* 83(1): 78–97 (2017), doi:10.1111/theo.12109.

Chapter 5

1. I'd like to say more, but the best evidence I can find are surveys of college students in rich countries, which is not a representative population. For example, Neil Dagnall, Kenneth Drinkwater, Andrew Parker, and Peter Clough, "Paranormal Experience, Belief in the Paranormal and Anomalous Beliefs," *Paranthropology: Journal of Anthropological Approaches to the Paranormal* 7(1): pp. 4–15 (2016).

2. Brian A. Sharpless and Karl Doghramji, *Sleep Paralysis: Historical, Psychological, and Medical Perspectives* (New York: Oxford University Press, 2015).

3. David J. Hufford, *The Terror That Comes in the Night: An Experience-Centered Study of Supernatural Assault Traditions* (Philadelphia: University of Pennsylvania Press, 1982), has a useful folkloristic description of the experience, despite his reluctance to endorse a straightforwardly physiological explanation.

4. Recent neuroscientific trends favor a model of the brain as an inference engine acting to minimize prediction error; see, for example, Anil K. Seth and Karl J. Friston, "Active Interoceptive Inference and the Emotional Brain," *Philosophical Transactions of the Royal Society B* 371: 20160007 (2016), doi:10.1098/rstb.2016.0007. Hallucinations might be better understood in such a context: Philip R. Corlett et al., "Hallucinations and Strong Priors," *Trends in Cognitive Sciences* 23(2): 114–127 (2019), doi:10.1016/j.tics.2018.12.001.

5. Scott O. Lilienfeld, Steven Jay Lynn, and Jeffrey M. Lohr, eds., *Science and Pseudoscience in Clinical Psychology* (2nd edition, New York: The Guilford Press, 2015).

6. Maarten Boudry, Fabio Paglieri, and Massimo Pigliucci, "The Fake, the Flimsy, and the Fallacious: Demarcating Arguments in Real Life," *Argumentation* 29: 431–456 (2015), doi:10.1007/s10503-015-9359-1. Maarten Boudry, "The Fallacy Fork: Why It's Time to Get Rid of Fallacy Theory," *Skeptical Inqui*rer 41(5): 46–51 (2017).

7. Christopher C. French and Anna Stone, *Anomalistic Psychology: Exploring Paranormal Belief and Experience* (New York: Palgrave Macmillan, 2014).

8. Jan-Willem van Prooijen, Karen M. Douglas, and Clara De Inocencio, "Connecting the Dots: Illusory Pattern Perception Predicts Belief in Conspiracies and the Supernatural," *European Journal of Social Psychology* 48: 320–335 (2018), doi:10.1002/ejsp.2331. Aiyana K. Willard and Ara Norenzayan, "Cognitive Biases Explain Religious Belief, Paranormal Belief, and Belief in Life's Purpose," *Cognition* 129(2): 379–391 (2013), doi:10.1016/j.cognition.2013.07.016. R. Kelly Garrett and Brian E. Weeks, "Epistemic Beliefs' Role in Promoting Misperceptions and Conspiracist Ideation," *PLoS ONE* 12(9): e0184733 (2017), doi:10.1371/journal.pone.0184733.

9. David M. Jacobs, *Walking Among Us: The Alien Plan to Control Humanity* (San Francisco: Disinformation Books, 2015).

10. I first encountered such ideas in Stewart Elliott Guthrie, *Faces in the Clouds: A New Theory of Religion* (New York: Oxford University Press, 1993). The cognitive science of religion then debated similar proposals, especially the "Hyperactive Agency Detection Device." Current reviews of the field are included in Jonathan H. Turner, Alexandra Maryanski, Anders Klostergaard Petersen, and Armin W. Geertz, eds., *The Emergence and Evolution of Religion: By Means of Natural Selection* (Milton Park: Taylor & Francis, 2017). Hans Van Eyghen and Konrad Szocik, *Revising Cognitive and Evolutionary Science of Religion: Religion as an Adaptation* (Cham: Springer Nature, 2021).

11. Pascal Boyer, *Religion Explained: The Evolutionary Origins of Religious*

Thought (New York: Basic Books, 2001). Pascal Boyer, *Minds Make Societies: How Cognition Explains the World Humans Create* (New Haven: Yale University Press, 2018), chapter 3.

12. Marc Andersen, "Predictive Coding in Agency Detection," *Religion, Brain & Behavior* 9(1): 65–84 (2019), doi:10.1080/2153599X.2017.1387170.

13. Dan Sperber et al., "Epistemic Vigilance," *Mind & Language* 25: 359–393 (2010), doi:10.1111/j.1468-0017.2010.01394.x. Hugo Mercier, *Not Born Yesterday: The Science of Who We Trust and What We Believe* (Princeton: Princeton University press, 2020).

14. For criticism, see Maria Kronfeldner, *Darwinian Creativity and Memetics* (Durham: Acumen, 2011). See, however, Maarten Boudry and Steije Hofhuis, "Parasites of the Mind. Why Cultural Theorists Need the Meme's Eye View," *Cognitive Systems Research* 52: 155–167 (2018), doi:10.1016/j.cogsys.2018.06.010.

15. Nicolas Claidière, Thomas C. Scott-Phillips, and Dan Sperber, "How Darwinian Is Cultural Evolution?" *Philosophical Transactions of the Royal Society B* 369: 20130368 (2014), doi:10.1098/rstb.2013.0368. Pascal Boyer, *Minds Make Societies: How Cognition Explains the World Humans Create* (New Haven: Yale University Press, 2018), 245–268.

16. Scott Atran and Joseph Henrich, "The Evolution of Religion: How Cognitive By-Products, Adaptive Learning Heuristics, Ritual Displays, and Group Competition Generate Deep Commitments to Prosocial Religions," *Biological Theory* 5(1): 18–30 (2010), doi:10.1162/BIOT_a_00018. Michael Vlerick, "The Cultural Evolution of Institutional Religions," *Religion, Brain & Behavior* 10(1): 18–34 (2020), doi:10.1080/2153599X.2018.1515105. Some theorists try to explain much of religion through prosocial features; for example, Ara Norenzayan, *Big Gods: How Religion Transformed Cooperation and Conflict* (Princeton: Princeton University Press, 2013). That emphasis, however, is questionable: Harvey Whitehouse et al., "Complex Societies Precede Moralizing Gods Throughout World History," *Nature* 568: 226–229 (2019), doi:10.1038/s41586-019-1043-4.

17. Economists have a history of adopting overly simplified notions of rationality, but instrumental rationality is still a good starting point. Jonathan Aldred, *Licence to be Bad: How Economics Corrupted Us* (London: Allen Lane, 2019).

18. I don't fully trust the rational choice program in the social sciences, and trust public choice theory even less, but these questions have most often been debated in such contexts. André Blais, *To Vote Or Not to Vote? The Merits and Limits of Rational Choice Theory* (Pittsburgh: University of Pittsburgh Press, 2000); Jonathan Bendor, Daniel Diermeier, David A. Siegel,

and Michael Ting, *A Behavioral Theory of Elections* (Princeton: Princeton University Press, 2011). On politicians ignoring low income voters, see Larry M. Bartels, *Unequal Democracy: The Political Economy of the New Gilded Age* (2nd edition, Princeton: Princeton University Press, 2016), 233–268.

19. "Progressive neoliberalism" also captures the flavor of liberal politics prevalent in the United States: Nancy Fraser, *The Old Is Dying and the New Cannot Be Born: From Progressive Neoliberalism to Trump and Beyond* (New York: Verso, 2019). I prefer to emphasize the meritocratic aspect of the politics of educated professionals. Daniel Markovitz, *The Meritocracy Trap: How America's Foundational Myth Feeds Inequality, Dismantles the Middle Class, and Devours the Elite* (New York: Penguin Press, 2019).

20. Some critics of anti-vaccine belief emphasize reasoning errors: Jonathan Howard and Dorit Rubinstein Reiss, "The Anti-Vaccine Movement: A Litany of Fallacy and Errors," in Allison B. Kaufman and James C. Kaufman, eds., *Pseudoscience: The Conspiracy Against Science* (Cambridge, MA: The MIT Press, 2018), 195–219. I am not convinced that producing laundry lists of alleged fallacies is an effective way of demonstrating reasoning errors, but I won't disagree that anti-vaccination notions are irrational.

21. Jennifer Abbasi, "Barry Marshall, MD: *H pylori* 35 Years Later," *JAMA* 317(14): 1400–1402 (2017), doi:10.1001/jama.2017.2629.

22. Ryan T. McKay and Daniel C. Dennett, "The Evolution of Misbelief," *Behavioral and Brain Sciences* 32(6): 493–561 (2009), doi:10.1017/S0140525X09990975.

23. Part of this chapter is based on Taner Edis and Maarten Boudry, "Truth and Consequences: When Is It Rational to Accept Falsehoods?" *Journal of Cognition and Culture* 19: 153–175 (2019); doi:10.1163/15685373-12340052. See also Konrad Talmont-Kaminski, *Religion as Magical Ideology: How the Supernatural Reflects Rationality* (Durham: Acumen, 2013), chapter 5.

24. While fictional, my Turania-Urartia example draws on the history of genocide and ethnic cleansing in the decline and aftermath of the Ottoman Empire, and the subsequent nationalisms in its successor states. For example, Benny Morris and Dror Ze'evi, *The Thirty-Year Genocide: Turkey's Destruction of Its Christian Minorities, 1894–1924* (Cambridge: Harvard University Press, 2019).

25. Real examples include pressure on American historians concerning controversies such as the Enola Gay, the reaction to historians who acknowledge the Armenian genocide in Turkey, and Hindu nationalist campaigns to teach their version of history. Edward T. Linenthal and Tom Engelhardt, *History Wars: The Enola Gay and Other Battles for the American Past* (New York: Henry Holt and Company, 1996); Taner Akçam, *The Young Turks'*

Crime Against Humanity: The Armenian Genocide and Ethnic Cleansing in the Ottoman Empire (Princeton: Princeton University Press, 2012); Sylvie Guichard, *The Construction of History and Nationalism in India Textbooks, Controversies and Politics* (New York: Routledge, 2010).

26. William von Hippel and Robert Trivers, "The Evolution and Psychology of Self-deception," *Behavioral and Brain Sciences* 34(1): 1–16 (2011), doi:10.1017/S0140525X10001354. Catherine Dunphy, *From Apostle to Apostate: The Story of the Clergy Project* (Durham: Pitchstone Publishing, 2015).

27. It is a costly, hard to fake signal. Joseph Bulbulia and Richard Sosis, "Signalling Theory and the Evolution of Religious Cooperation," *Religion* 41(3): 363–388 (2011), doi:10.1080/0048721X.2011.604508.

28. Dan M. Kahan, "The Politically Motivated Reasoning Paradigm, Part 1: What Politically Motivated Reasoning Is and How to Measure It," in Robert A. Scott and Stephen Michael Kosslyn, eds., *Emerging Trends in the Social and Behavioral Sciences* (Wiley Online Library, 2016), doi:10.1002/9781118900772.etrds0417. Daniel Williams, "Socially Adaptive Belief," *Mind & Language* 1–22 (2020), doi:10.1111/mila.12294.

29. Conventional decision theory describes a form of instrumental rationality where the interests to be satisfied are fixed and external to the decision process. Martin Peterson, *An Introduction to Decision Theory* (2nd edition, New York: Cambridge University Press, 2017).

30. Richard D. Alexander, *The Biology of Moral Systems* (New Brunswick: AldineTransaction, 1987). Mikael Klintman, *Human Sciences and Human Interests: Integrating the Social, Economic, and Evolutionary Sciences* (New York: Routledge, 2017).

31. Miguel A. Altieri, Clara I. Nicholls, and Rene Montalba, "Technological Approaches to Sustainable Agriculture at a Crossroads: An Agroecological Perspective," *Sustainability* 9(3): 349 (2017), doi:10.3390/su9030349.

32. David Sehat, *The Myth of American Religious Freedom* (New York: Oxford University Press, 2011).

33. Maarten Boudry and Johan Braeckman. "How Convenient! The Epistemic Rationale of Self-validating Belief Systems," *Philosophical Psychology* 25(3): 341–364 (2012). doi:10.1080/09515089.2011.579420.

34. Marc A. Edwards and Siddhartha Roy, "Academic Research in the 21st Century: Maintaining Scientific Integrity in a Climate of Perverse Incentives and Hypercompetition," *Environmental Engineering Science* 34(1): 51–61 (2017), doi:10.1089/ees.2016.0223. Stuart Ritchie, *Science Fictions: How Fraud, Bias, Negligence, and Hype Undermine the Search for Truth* (New York: Metropolitan Books, 2020), 175–237.

35. Joel Spring, *Economization of Education: Human Capital, Global Corpora-tions, Skills-Based Schooling* (New York: Routledge, 2015); Henry A. Gir-oux, *Neoliberalism's War on Higher Education* (Chicago: Haymarket Books, 2014). At all levels, science education has become increasingly justified only in economic terms; John L. Rudolph, *How We Teach Science: What's Changed, and Why It Matters* (Cambridge: Harvard University Press, 2019).

36. Taner Edis, "A Revolt Against Expertise: Pseudoscience, Right-Wing Popu-lism, and Post-Truth Politics," *Disputatio Philosophical Research Bulletin* 9(13) (2020), doi:10.5281/zenodo.3567166. Harry Collins, Robert Evans, Darrin Durant, and Martin Weinel, *Experts and the Will of the People: Soci-ety, Populism and Science* (Cham: Springer Nature, 2019).

37. I am describing some common intuitions that lead into varieties of moral realism mostly familiar in Western, modern, and religious contexts. Widen-ing the focus in time and space leads to other options, which I don't consid-er: Colin Marshall, ed., *Comparative Metaethics: Neglected Perspectives on the Foundations of Morality* (New York: Routledge, 2020). Further variety, I think, favors a non-realist view. Also, in populations beyond philosophers, there is considerable metaethical variety and a wider range of intuitions than those I address: Jennifer Cole Wright, "The Fact and Function of Me-ta-Ethical Pluralism: Exploring the Evidence," in Tania Lombrozo, Joshua Knobe, and Shaun Nichols, eds., *Oxford Studies in Experimental Philosophy, Volume 2* (New York: Oxford University Press, 2018). My target is only the cluster of intuitions that support realism.

38. I favor an amoralist metaethics: Russell Blackford, *The Mystery of Moral Authority* (New York: Palgrave Macmillan, 2016); Joel Marks, *Ethics With-out Morals: In Defense of Amorality* (New York: Routledge, 2013). More generally, however, I have no objection to subjectivist metaethics such as Richard Double, *Metaethical Subjectivism* (New York: Routledge, 2006). I just don't think there are moral facts, including moral facts supervening on facts about the natural world. Taner Edis, "Two Cheers for Scientism," in Maarten Boudry and Massimo Pigliucci, eds., *Science Unlimited? The Chal-lenges of Scientism* (Chicago: University of Chicago Press, 2018).

39. More generally, reasoning is deeply involved in our moral psychology. Han-no Sauer, *Moral Judgments as Educated Intuitions* (Cambridge: The MIT Press, 2017); Michael Vlerick, "Explaining Human Altruism," *Synthese* (2020), doi:10.1007/s11229-020-02890-y. Morality cannot be collapsed onto mere preferences, as some economists assume; nonetheless, moral reasoning does not refer to anything beyond ordinary physical reality.

40. Beyond world religions and their official theologies, the actual relationship between religion and morality is far more complicated. Ryan McKay and

Harvey Whitehouse, "Religion and Morality," *Psychological Bulletin*, 141(2): 447–473 (2015), doi:10.1037/a0038455.

41. Philosophers and psychologists who study moral behavior often focus on psychopaths as an extreme case to test moral theories; Thomas Schramme, ed., *Being Amoral: Psychopathy and Moral Incapacity* (Cambridge: The MIT press, 2014). I think a more telling example is not a psychopath but a "classist amoralist" as imagined by Kai Nielsen, *Why Be Moral?* (Amherst: Prometheus Books, 1989), 295–298.

42. Will M. Gervais, et al., "Global evidence of extreme intuitive moral prejudice against atheists," *Nature Human Behaviour* 1: 0151 (2017), doi:10.1038/s41562-017-0151.

43. Some atheists who hope to replace religion with a secular alternative provide the best examples. Richard Carrier, "Moral Facts Naturally Exist (and Science Could Find Them)," in John W. Loftus, ed., *The End of Christianity* (Amherst, NY: Prometheus Books, 2011). Sam Harris, *The Moral Landscape: How Science can Determine Human Values* (New York: Free Press, 2010).

44. Even if human nature were to underwrite broad commonalities in what humans value, there need be no overarching ordering principle to adjudicate such value conflicts. Michael B. Gill, *Humean Moral Pluralism* (Oxford: Oxford University Press, 2014). In a political context, this manifests as "value pluralism"; Peter Lassman, *Pluralism* (Malden: Polity Press, 2011).

45. Taner Edis, *Islam Evolving: Radicalism, Reformation, and the Uneasy Relationship with the Secular West* (Amherst: Prometheus Books, 2016), 238–246.

46. Debates over human enhancement implicitly involve such questions, but in what I've read, metaethical concerns usually get swamped by normative arguments. Steve Clarke, Julian Savulescu, C. A. J. Coady, Alberto Giubilini, and Sagar Sanyal, eds., *The Ethics of Human Enhancement: Understanding the Debate* (New York: Oxford University Press, 2016).

47. David Owens, "Disenchantment," in Louise Anthony, ed., *Philosophers without Gods: Meditations on Atheism and the Secular Life* (New York: Oxford University Press, 2007).

48. Donald R. Prothero, *Reality Check: How Science Deniers Threaten Our Future* (Bloomington: Indiana University Press, 2013).

49. Taner Edis and Amy S. Bix, "Biology and 'Created Nature': Gender and the Body in Popular Islamic Literature from Modern Turkey and the West," *Arab Studies Journal* 12(2)/13(1): 140–58 (2005).

50. Sonja Brentjes, Taner Edis, and Lutz Richter-Bernburg, eds., *1001 Distortions: How (Not) to Narrate History of Science, Medicine, and Technology in*

Non-Western Cultures (Würtzburg: Ergon-Verlag, 2016).

51. Peter Morey, Amina Yaqin, and Alaya Forte, eds., *Contesting Islamophobia: Anti-Muslim Prejudice in Media, Culture and Politics* (London: Bloomsbury Academic, 2019).

52. Quoted in Jennifer Simkins, *The Science Fiction Mythmakers: Religion, Science and Philosophy in Wells, Clarke, Dick and Herbert* (Jefferson: McFarland & Company, 2016), 72.

53. Hubert Dreyfus and Sean Dorrance Kelly, *All Things Shining: Reading the Western Classics to Find Meaning in a Secular Age* (New York: Free Press, 2011).

54. For example, Pamela Geller, "Pamela Geller, Breitbart News: '1001 Muslim Myths and Historical Revisions,'" *Geller Report*, July 26, 2015, https://gellerreport.com/2015/07/pamela-geller-breitbart-news-1001-muslim-myths-and-historical-revisions.html/, accessed January 9, 2021.

55. Robert Boyers, *The Tyranny of Virtue: Identity, the Academy, and the Hunt for Political Heresies* (New York: Scribner, 2019).

Chapter 6

1. Richard Dawid, *String Theory and the Scientific Method* (Cambridge: Cambridge University Press, 2013). Radin Dardashti, Richard Dawid, and Karim Thébault, eds., *Why Trust a Theory? Epistemology of Fundamental Physics* (Cambridge: Cambridge University Press, 2019).

2. Toby Ord, *The Precipice: Existential Risk and the Future of Humanity* (New York: Hachette Books, 2020). Depressingly, even human expansion into space, a dream of many scientists and science-fiction fans, is an existential threat. Daniel Deudny, *Dark Skies: Space Expansionism, Planetary Geopolitics, and the Ends of Humanity* (New York: Oxford University Press, 2020).

3. Kendrick Frazier, "History of CSICOP," in Gordon Stein, ed., *The Encyclopedia of the Paranormal* (Amherst: Prometheus Books, 1996), 168-180.

4. Rebekah Higgitt, "Is There a Rising Tide of Irrationality?" *Guardian*, November 21, 2012, https://www.theguardian.com/science/the-h-word/2012/nov/21/history-science, accessed January 15, 2021. Matt Nisbet, "Cultural Indicators Of The Paranormal," *Skeptical Inquirer* (web) https://skepticalinquirer.org/exclusive/cultural-indicators-of-the-paranormal/ (2006), accessed January 15, 2021. Naomi Oreskes and Erik M. Conway, *Merchants of Doubt: How a Handful of Scientists Obscured the Truth on Issues From Tobacco Smoke to Global Warming* (New York: Bloomsbury Press, 2010). Núria Almiron, Maxwell Boykoff, Marta Narberhaus, and Francisco Heras, "Dominant Counter-frames in Influential Climate Contrarian European

Think Tanks," *Climatic Change* 162(4): 2003–2020 (2020), doi:10.1007/s10584-020-02820-4.

5. Parts of this chapter are based on Taner Edis, "A Revolt Against Expertise: Pseudoscience, Right-Wing Populism, and Post-Truth Politics," *Disputatio Philosophical Research Bulletin* 9:13 (2020), doi:10.5281/zenodo.3567166.

6. For example, Gleb Tsipursky and Tim Ward, *Pro Truth: A Practical Plan for Putting Truth Back into Politics* (Washington: Changemakers Books, 2020).

7. Steve Bruce, *Secular Beats Spiritual: The Westernization of the Easternization of the West* (New York: Oxford University Press, 2017). The long-term stability of secularity, however, can be doubted; Wesley J. Wildman, F. LeRon Shults, Saikou Y. Diallo, Ross Gore, and Justin Lane, "Post-Supernatural Cultures: There and Back Again," *Secularism and Nonreligion* 9(6): 1–15 (2020), doi:10.5334/snr.121.

8. Deborah Kelemen, "Teleological Minds: How Natural Intuitions about Agency and Purpose Influence Learning about Evolution," in Karl S. Rosengren, Sarah K. Brem, E. Margaret Evans, and Gale M. Sinatra, eds., *Evolution Challenges: Integrating Research and Practice in Teaching and Learning about Evolution* (New York: Oxford University Press, 2012).

9. Ronald L. Numbers, *The Creationists: From Scientific Creationism to Intelligent Design* (Cambridge: Harvard University Press, 2006). The role of competition and conflict *within* conservative Christianity in the development of the varieties of American creationism is also interesting: Benjamin L. Huskinson, *American Creationism, Creation Science, and Intelligent Design in the Evangelical Market* (Cham: Springer Nature, 2020).

10. Henry M. Morris, *The Long War Against God: The History and Impact of the Creation/Evolution Conflict* (Green Forest: Master Books, 2000). Scott S. Powell, "The Deep State Digs Deeper," *Discovery Institute*, January 16, 2019, https://www.discovery.org/a/21435/, accessed January 15, 2021. Taner Edis, "Cosmic Conspiracy Theories: How Theologies Evade Science," in Joseph Seckbach and Richard Gordon, eds., *Theology and Science: From Genesis to Astrobiology* (Singapore: World Scientific, 2019).

11. Robin Globus Veldman, *The Gospel of Climate Skepticism: Why Evangelical Christians Oppose Action on Climate Change* (Oakland: University of California Press, 2019). Bernard Daley Zaleha and Andrew Szasz, "Why Conservative Christians Don't Believe in Climate Change," *Bulletin of the Atomic Scientists* 71(5): 19–30 (2015), doi:10.1177/0096340215599789.

12. Julie J. Ingersoll, *Building God's Kingdom: Inside the World of Christian Reconstruction* (New York: Oxford University Press, 2015).

13. I elaborate my descriptions of Turkish creationism in Taner Edis, "The Politics of Islamic Opposition to Evolution in Turkey," in C. Mackenzie Brown,

ed., *Asian Religious Responses to Darwinism: Evolutionary Theories in Middle Eastern, South Asian, and East Asian Cultural Contexts* (Cham: Springer Nature, 2020); Taner Edis, "The Turkish Model of Islamic Creationism," *Almagest* 12: 40–65 (2021), doi:10.1484/J.ALMAGEST.5.125385.

14. Robert M. Geraci, *Temples of Modernity: Nationalism, Hinduism, and Transhumanism in South Indian Science* (Lanham: Lexington Books, 2018). Meera Nanda, *Prophets Facing Backward: Postmodern Critiques of Science and Hindu Nationalism in India* (Piscataway: Rutgers University Press, 2003). Priya Chacko, "The Right Turn in India: Authoritarianism, Populism and Neoliberalisation," *Journal of Contemporary Asia* 48(4): 541–565 (2018), doi:10.1080/00472336.2018.1446546. Priyanka Pulla, "'A Fraud on the Nation': Critics Blast Indian Government's Promotion of Traditional Medicine for COVID-19," *Science*, October 15, 2020, doi:10.1126/science.abf2671.

15. For an example of creationist rhetoric of "discernment" and the Bible as a criterion of truth, see John C. P. Smith, "Truth about Origins and Faith," *Answers in Genesis*, May 1, 2015, https://answersingenesis.org/hermeneutics/truth-about-origins-and-faith/, accessed January 15, 2021. For evolution as the preoccupation of a liberal elite, see Frank S. Smith, *The Liberal Manifesto: A Concise Introduction to the Unsustainable Ideology of Liberalism* (Arlington: Perspectus Publishing, 2012).

16. "Science literacy" tends to be hard to define and includes too many demands. Catherine E. Snow and Kenne A. Dibner, eds., *Science Literacy: Concepts, Contexts, and Consequences* (Washington: The National Academies Press, 2016). I want science to be *available* to all, but also agree with critics who argue that mass substantial science literacy is an unrealistic goal. Morris H. Shamos, *The Myth of Scientific Literacy* (New Brunswick: Rutgers University Press, 1995).

17. Research tends to address paranormal and religious beliefs in general. James A. Wilson, "Reducing Pseudoscientific and Paranormal Beliefs in University Students Through a Course in Science and Critical Thinking," *Science & Education* 27: 183–210 (2018), doi:10.1007/s11191-018-9956-0. Christopher C. French and Anna Stone, *Anomalistic Psychology: Exploring Paranormal Belief and Experience* (New York: Palgrave Macmillan, 2014), 31, 143–144.

18. Jerry A. Coyne, "Science, Religion, and Society: The Problem of Evolution in America," *Evolution* 66: 2654–2663 (2012), doi:10.1111/j.1558-5646.2012.01664.x.

19. "Evolution, Creationism, Intelligent Design," *Gallup*, https://news.gallup.com/poll/21814/Evolution-Creationism-Intelligent-Design.aspx, accessed

December 31, 2020. On the secularization of the United States, see Kevin McCaffree, *The Secular Landscape: The Decline of Religion in America* (New York: Palgrave Macmillan, 2017); Landon Schnabel and Sean Bock, "The Persistent and Exceptional Intensity of American Religion: A Response to Recent Research," *Sociological Science* 5: 711–721 (2018), doi:10.15195/v4.a28.

20. Taner Edis, "Is There A Political Argument For Teaching Evolution?" *Marburg Journal of Religion* 22:2 (2020), doi:10.17192/mjr.2020.22.8304.

21. Daniel Markovitz, *The Meritocracy Trap: How America's Foundational Myth Feeds Inequality, Dismantles the Middle Class, and Devours the Elite* (New York: Penguin Press, 2019). Jo Littler, *Against Meritocracy: Culture, Power and Myths of Mobility* (New York: Routledge, 2017). Mark Bovens and Anchrit Wille, *Diploma Democracy: The Rise of Political Meritocracy* (New York: Oxford University Press, 2017). Jochem van Noord, Bram Spruyt, Toon Kuppens, and Russell Spears, "Education-Based Status in Comparative Perspective: The Legitimization of Education as a Basis for Social Stratification," *Social Forces* 98(2): 649–676 (2019), doi:10.1093/sf/soz012.

22. Lawrence Baum, *The Supreme Court* (13th edition, Thousand Oaks: CQ Press, 2018), 183–185.

23. Taner Edis, "Technological Progress and Pious Modernity: Secular Liberals Fall Behind the Times," in Anthony B. Pinn, ed., *Humanism and Technology: Opportunities and Challenges* (New York: Palgrave Macmillan, 2016).

24. For example, Michiko Kakutani, *The Death of Truth: Notes on Falsehood in the Age of Trump* (New York: Tim Duggan Books, 2018). For criticism of Russia conspiracy theories, see Alan Macleod, "Fake News, Russian Bots and Putin's Puppets," in Alan Macleod, ed., *Propaganda in the Information Age: Still Manufacturing Consent* (New York: Routledge, 2019).

25. Such criticisms are often associated with the political left and the charge that mainstream media are permeated by propaganda; Joan Pedro-Carañana, Daniel Broudy, and Jeffery Klaehn, *The Propaganda Model Today: Filtering Perception and Awareness* (London: University of Westminster Press, 2018); Alan Macleod, ed., *Propaganda in the Information Age: Still Manufacturing Consent* (New York: Routledge, 2019). To me, these criticisms ring true. Especially with the overt partisanship of the US media in the Trump era, social conservatives have also become more aware of problems such as media bias. Jim A. Kuypers, *President Trump and the News Media: Moral Foundations, Framing, and the Nature of Press Bias in America* (Lanham: Lexington Books, 2020).

26. Musa al Gharbi, "Race and the Race for the White House: On Social Research in the Age of Trump," *The American Sociologist* 49: 496–519 (2018),

doi:10.1007/s12108-018-9373-5.

27. Pippa Norris and Ronald Inglehart, *Sacred and Secular: Religion and Politics Worldwide* (New York: Cambridge University Press, 2011). Pippa Norris and Ronald Inglehart, *Cultural Backlash: Trump, Brexit, and Authoritarian Populism* (New York: Cambridge University Press, 2019). Political theorists have some explanations; for example, Wendy Brown, *In the Ruins of Neoliberalism: The Rise of Antidemocratic Politics in the West* (New York: Columbia University Press, 2019), but Brown overlooks the element of reaction to meritocracy in populism.

28. Arlie Russell Hochschild, *Strangers in Their Own Land: Anger and Mourning on the American Right* (New York: New Press, 2016). Jacob S. Hacker and Paul Pierson, *Let Them Eat Tweets: How the Right Rules in an Age of Extreme Inequality* (New York: Liveright, 2020). Right-wing populist movements often have business connections; Neşecan Balkan, Erol Balkan, and Ahmet Öncü, eds., *Neoliberalizm, İslamcı Sermayenin Yükselişi ve AKP* (İstanbul: Yordam Kitap, 2013); Kevin M. Kruse, *One Nation Under God: How Corporate America Invented Christian America* (New York: Basic Books, 2015).

29. Taner Edis, *Islam Evolving: Radicalism, Reformation, and the Uneasy Relationship with the Secular West* (Amherst: Prometheus Books, 2016).

30. Globally, there is plenty of middle-class opposition to even formal democracy. Joshua Kurlantzick, *Democracy in Retreat: The Revolt of the Middle Class and the Worldwide Decline of Representative Government* (New Haven: Yale University Press, 2013). Historically, democratic participation has advanced due to the demands of non-elite and working class actors; Adaner Usmani, "Democracy and the Class Struggle," *American Journal of Sociology* 124(3): 664–704 (2018), doi:10.1086/700235. On anti-populist and anti-democratic politics in the US, see Thomas Frank, *The People, No: A Brief History of Anti-Populism* (New York: Metropolitan Books, 2020).

31. Martin Gurri, *The Revolt of the Public and the Crisis of Authority in the New Millennium* (San Francisco: Stripe Press, 2018). Yochai Benkler, Robert Faris, and Hal Roberts, *Network Propaganda: Manipulation, Disinformation, and Radicalization in American Politics* (New York: Oxford University Press, 2018). Mark Davis, "The Online Anti-Public Sphere," *European Journal of Cultural Studies* (2020), doi:10.1177/1367549420902799.

32. There is some complexity in far-right attitudes toward the environment, even including "ecofascist" strains. But by and large, right-wing populism is associated with anti-environmental views. Bernhard Forchtner, *The Far Right and the Environment: Politics, Discourse and Communication* (New York: Routledge, 2020). Ville-Juhani Ilmarinen, Florencia M.Sortheix, and Jan-Erik Lönnqvist, "Consistency and Variation in the Associations be-

tween Refugee and Environmental Attitudes in European Mass Publics," *Journal of Environmental Psychology* 73: 101540 (2021), doi:10.1016/j.jenvp.2020.101540.

33. James K. Galbraith, "The Predator State," *Catalyst* 1(3) (2017), https://catalyst- journal.com/vol1/no3/the-predator-state, accessed January 16, 2021.

34. Dorien Zandbergen, "Silicon Valley New Age: The Co-Constitution Of The Digital And The Sacred," in Stef Aupers and Dick Houtman, eds., *Religions of Modernity: Relocating the Sacred to the Self and the Digital* (Dordrecht: Brill, 2010), doi:10.1163/ej.9789004184510.i-273.61. James Dennis LoRusso, *Spirituality, Corporate Culture, and American Business: The Neoliberal Ethic and the Spirit of Global Capital* (New York: Bloomsbury Academic, 2017). An interesting extreme that goes beyond mushy spirituality and libertarianism is described by Alexander J. Means and Graham B. Slater, "The Dark Mirror of Capital: On Post-Neoliberal Formations and the Future of Education," *Discourse: Studies in the Cultural Politics of Education* 40(2): 162–175 (2019), doi:10.1080/01596306.2019.1569876.

35. Evgeny Morozov, *To Save Everything, Click Here: The Folly of Technological Solutionism* (New York: Public Affairs, 2013). Philanthropy is permeated by a similar search for technical solutions that don't disrupt the status quo. Anand Giridharadas, *Winners take All: The Elite Charade of Changing the World* (New York: Alfred A. Knopf, 2018). On innovation, see Mariana Mazzucato, *The Entrepreneurial State: Debunking Public vs. Private Sector Myths* (revised edition, London: Anthem, 2015).

36. Newley Purnell and Jeff Horwitz, "Facebook's Hate-Speech Rules Collide With Indian Politics," *The Wall Street Journal*, Aug. 14, 2020, https://www.wsj.com/articles/facebook-hate-speech-india-politics-muslim-hindu-modi-zuckerberg-11597423346, accessed January 16, 2021. Alan Macleod, "That Facebook Will Turn to Censoring the Left Isn't a Worry—It's a Reality," *FAIR*, August 22, 2018, https://fair.org/home/that-facebook-will-turn-to-censoring-the-left-isnt-a-worry-its-a-reality/, accessed January 16, 2021.

37. Damian J. Ruck, R. Alexander Bentley, and Daniel J. Lawson, "Cultural Prerequisites of Socioeconomic Development," *Royal Society Open Science* 7: 190725 (2020), doi:10.1098/rsos.190725.

38. Domenico Losurdo, *Liberalism: A Counter-History* (New York: Verso, 2011). Mike Davis, *Late Victorian Holocausts: El Niño Famines and the Making of the Third World* (New York: Verso, 2001).

39. Steven Pinker, *Enlightenment Now: The Case for Reason, Science, Humanism, and Progress* (New York: Penguin Books, 2018). Hans Rosling, Anna Rosling Rönnlund, and Ola Rosling, *Factfulness: Ten Reasons We're Wrong*

About the World—and Why Things Are Better Than You Think (New York: Flatiron Books, 2018). For criticism, see Jeremy Lent, "Steven Pinker's Ideas are Fatally Flawed. These Eight Graphs Show Why," *OpenDemocracy*, May 21, 2018, https://www.opendemocracy.net/en/transformation/steven-pinker-s-ideas-are-fatally-flawed-these-eight-graphs-show-why/, accessed January 17, 2021.

40. Christopher Ryan, *Civilized to Death: The Price of Progress* (New York: Avid Reader Press, 2019). James C. Scott, *Against the Grain: A Deep History of the Earliest States* (New Haven: Yale University Press, 2017).

41. Jason Hickel, *The Divide: A Brief Guide to Global Inequality and its Solutions* (New York: W. W. Norton, 2018). Jason Hickel, "Is Global Inequality Getting Better or Worse? A Critique of the World Bank's Convergence Narrative," *Third World Quarterly* 38(10): 2208–2222 (2017), doi:10.1080/01436 597.2017.1333414. Jason Hickel, "A Letter to Steven Pinker (and Bill Gates, for that matter) About Global Poverty," *Class, Race and Corporate Power* 7(1): 3 (2019), doi:10.25148/CRCP.7.1.008318. Philip Alston, "The Parlous State of Poverty Eradication: Report of the Special Rapporteur on Extreme Poverty and Human Rights," July 2, 2020, https://chrgj.org/wp-content/uploads/2020/07/Alston-Poverty-Report-FINAL.pdf, accessed January 17, 2021. On India, see essays in Arundhati Roy, *My Seditious Heart: Collected Nonfiction* (Chicago: Haymarket Books, 2019).

42. Suzana Sawyer and Edmund Terence Gomez, *The Politics of Resource Extraction: Indigenous Peoples, Multinational Corporations and the State* (New York: Palgrave Macmillan, 2012). James Suzman, *Affluence Without Abundance: The Disappearing World of the Bushmen* (New York: Bloomsbury Press, 2017).

43. Michaël Aklin, "Re-exploring the Trade and Environment Nexus Through the Diffusion of Pollution," *Environmental and Resource Economics* 64: 663–682 (2016), doi:10.1007/s10640-015-9893-1. Erik Loomis, *Out of Sight: The Long and Disturbing Story of Corporations Outsourcing Catastrophe* (New York: The New Press, 2015).

44. Jason Hickel and Giorgos Kallis, "Is Green Growth Possible?" *New Political Economy* 4: 469–486 (2019), doi:10.1080/13563467.2019.1598964. There has been much overpromising associated with environmentally friendlier energy technologies. Christopher T. M. Clack et al., "Evaluation of a Proposal for Reliable Low-Cost Grid Power with 100% Wind, Water, and Solar," *Proceedings of the National Academy of Sciences* 114(26): 6722–6727 (2017), doi:10.1073/pnas.1610381114. Carlos de Castro and Iñigo Capellán-Pérez, "Concentrated Solar Power: Actual Performance and Foreseeable Future in High Penetration Scenarios of Renewable Energies," *BioPhysical Economics*

and Resource Quality 3: 14 (2018), doi:10.1007/s41247-018-0043-6.

45. Steven Pinker, "Progressophobia: Why Things are Better Than You Think They Are," *Skeptical Inquirer* 42(3): 26–35 (2018). Links between skepticism and "new atheist" movements are also a reason to be cautious about entanglements with ideology. Stephen LeDrew, *The Evolution of Atheism: The Politics of a Modern Movement* (New York: Oxford University Press, 2016).

46. Corey J. A. Bradshaw et al., "Underestimating the Challenges of Avoiding a Ghastly Future," *Frontiers in Conservation Science* 1:615419 (2021), doi:10.3389/fcosc.2020.615419.

47. Keynyn Brysse, Naomi Oreskes, Jessica O'Reilly, and Michael Oppenheimer, "Climate Change Prediction: Erring on the Side of Least Drama?" *Global Environmental Change* 23 (1): 327–337 (2013), doi:10.1016/j.gloenvcha.2012.10.008.

48. Christopher Newfield, *The Great Mistake: How We Wrecked Public Universities and How We Can Fix Them* (Baltimore: The Johns Hopkins University Press, 2016). Herb Childress, The *Adjunct Underclass: How America's Colleges Betrayed Their Faculty, Their Students, and Their Mission* (Chicago: University of Chicago Press, 2019).

49. Marc A. Edwards and Siddhartha Roy, "Academic Research in the 21st Century: Maintaining Scientific Integrity in a Climate of Perverse Incentives and Hypercompetition," *Environmental Engineering Science* 34(1): 51–61 (2017), doi:10.1089/ees.2016.0223. Stuart Ritchie, *Science Fictions: How Fraud, Bias, Negligence, and Hype Undermine the Search for Truth* (New York: Metropolitan Books, 2020), 175–237.

50. Philip Mirowski, *Science-Mart: Privatizing American Science* (Cambridge: Harvard University Press, 2011). Raphael Sassower, *Compromising the Ideals of Science* (New York: Palgrave Macmillan, 2015). David Tyfield, Rebecca Lave, Samuel Randalls, and Charles Thorpe, eds., *The Routledge Handbook of the Political Economy of Science* (New York: Routledge, 2017).

51. The archives of the creation congresses: "Kongre Arşivi" (2020), http://yaratiliskongresi.dpu.edu.tr/arsiv.html, accessed January 17, 2021.

52. Maarten Boudry, Michael Vlerick, and Taner Edis, "The End of Science? On Human Cognitive Limitations and How to Overcome Them," *Biology & Philosophy* 35: 18 (2020); doi:10.1007/s10539-020-9734-7.

53. Economists, even those who seriously contemplate no-growth scenarios, continue to be far too sanguine about environmental limits, relying too much on future innovation. For example, Cameron Hepburn and Alex Bowen, "Prosperity With Growth: Economic Growth, Climate Change and Environmental Limits," in Roger Fouquet, ed., *Handbook of Energy and Cli-*

mate Change (Northampton: Edward Elgar Publishing, 2013), doi:10.4337/9780857933690.00041.

54. Anders Esmark, *The New Technocracy* (Bristol: Bristol University Press, 2020). Jeffrey Friedman, *Power Without Knowledge: A Critique of Technocracy* (New York: Oxford University Press, 2020).

55. Samuel Scheffler, *Death and the Afterlife* (New York: Oxford University Press, 2013), argues that the value of what we do depends on the continued existence of generations of humans to come. I don't have good answers to questions about what really matters or what is the meaning of life; 42, I suppose. But a collapse of civilization or human extinction certainly would be catastrophic for all I care about.

56. Taner Edis, *Islam Evolving: Radicalism, Reformation, and the Uneasy Relationship with the Secular West* (Amherst: Prometheus Books, 2016), chapter 6.

57. Duncan H. Forgan, *Solving Fermi's Paradox* (New York: Cambridge University Press, 2019). Nathan Alexander Sears, "Anarchy, Technology, and the Self-Destruction Hypothesis: Human Survival and the Fermi Paradox," *World Futures* 76(8): 579–610 (2020), doi:10.1080/02604027.2020.1819116.

INDEX

ABOUT THE AUTHOR

Taner Edis is a professor of physics at Truman State University. His primary research has been in the history and philosophy of science, addressing questions raised by popular beliefs in fake science and the paranormal. Accordingly, the sign on his office door reads, "Taner Edis, Physics and Weirdness." He has authored and co-edited many books, including *Why Intelligent Design Fails* and *The Ghost in the Universe*. He lives in Kirksville, Missouri.